WHAT PEOPLE ARE SAYING [illegible]

Neglected or M[illegible] dical Feminism o[illegible]

T0098101

Clear-eyed, pedagogical and contemporary, Victoria Margree's reading of Firestone will be of use to anyone thinking about feminism and technology today. Providing clear historical and philosophical context, Margree provides a supremely balanced account of Firestone in all her complexity. With reference to other great feminist thinkers, the women's political movement, as well as recent developments in reproductive technology, Margree's book brings Firestone firmly up-to-date.
Nina Power, author of *One Dimensional Woman*

Questioning, rigorous, yet accessible, this is an important contribution to the current reconsideration of the demonized visionary of Second Wave feminism. With careful attention to its intellectual influences and analytical coherence, Victoria Margree offers a lucid exposition of Shulamith Firestone's argument against the power imbalances created by maternity. As the conservative challenge to contraception and abortion intensifies in Europe and the US, Margree concludes with an invaluable discussion of the post-Firestone politics of reproduction.
Mandy Merck, Co-Editor, *Further Adventures of the Dialectic of Sex*

Neglected or Misunderstood: The Radical Feminism of Shulamith Firestone

Neglected or Misunderstood: The Radical Feminism of Shulamith Firestone

Victoria Margree

Winchester, UK
Washington, USA

First published by Zero Books, 2018
Zero Books is an imprint of John Hunt Publishing Ltd., No. 3 East Street
Alresford, Hampshire SO24 9EE, UK
office1@jhpbooks.net
www.johnhuntpublishing.com
www.zero-books.net

For distributor details and how to order please visit the 'Ordering' section on our website.

Text copyright: Victoria Margree 2017

ISBN: 978 1 78535 539 4
978 1 78535 540 0 (ebook)
Library of Congress Control Number: 2016962872

All rights reserved. Except for brief quotations in critical articles or reviews, no part of this book may be reproduced in any manner without prior written permission from the publishers.

The rights of Victoria Margree as author have been asserted in accordance with the Copyright, Designs and Patents Act 1988.

A CIP catalogue record for this book is available from the British Library.

Design: Stuart Davies

Printed and bound by CPI Group (UK) Ltd, Croydon, CR0 4YY, UK

We operate a distinctive and ethical publishing philosophy in all areas of our business, from our global network of authors to production and worldwide distribution.

Contents

To my parents, for loving us for our own sakes

Acknowledgments

This book comes out of ten years' experience of teaching Shulamith Firestone to undergraduates on the Humanities Programme at the University of Brighton. I have many people to thank individually. Samantha Goodyer first suggested the project to me, for which I am deeply grateful. I have enjoyed great conversations about Firestone with Joanne Tutt, David Martin and Gill Scott. Antonia Hofstaetter gave helpful feedback on some early material for the book and I have valued her interest and encouragement. In particular, I would like to thank Gurminder Bhambra, Bob Brecher and Michael Neu, for their support and friendship, and for their invaluable feedback and suggestions on drafts. All mistakes that remain are my own. And I thank my partner, Moritz Schick, for supporting me in the project even when I have no doubt become something of a Firestone bore. But I'd also like to thank the generations of students for the stimulating and revealing classroom conversations from which my own ability to think through the challenges posed by Firestone's work has benefited in incalculable ways.

1

Introduction: In Defense of Science Fiction

> *"How can men be mothers! How can some kid who isn't related to you be your child?"...How could anyone know what being a mother means who has never carried a child nine months heavy under her heart, who has never borne a baby in blood and pain, who has never suckled a child.*
>
> (Marge Piercy[1])

An inhabitant of 1970s New York, Connie Ramos has been telepathically transported to a future society in which pregnancy and childbirth have been eliminated. Children are grown from embryos in artificial wombs and are raised by multiple 'mothers' not genetically related to them, and who may be either men or women. Connie has responded with curiosity but by no means hostility to many aspects of this brave new world, but as the door of the 'brooder' tank slides back, revealing 'seven human babies joggling slowly upside down, each in a sac of its own inside a larger fluid receptacle,' her mouth gapes and her stomach turns.[2] That *biological motherhood itself* could be expropriated from women fills her with revulsion.

Yet this is no dystopian vision, but a utopian one. And despite Connie's fears about male encroachment upon the unique capacity of women, it is a feminist one. It is drawn from Marge Piercy's 1976 novel *Woman on the Edge of Time,* which imaginatively creates the future society sketched out in Shulamith Firestone's feminist manifesto, *The Dialectic of Sex: The Case for Feminist Revolution*.

Firestone's book burst onto the emerging feminist scene in 1970 and proved immediately controversial. Its main thesis is that the origin of women's oppression lies in biology itself. For

Firestone, it is precisely the fact that it is women and not men who gestate a child, give birth in blood and pain, and shape their lives around another dependent human being that gives rise to male domination. This biological difference, she argues, divides humanity into two classes that are not equal, and this fundamental inequality then reproduces itself remorselessly at all levels of society. If Firestone's analysis was stark, then her solution was revolutionary: since it is biology that is the problem, then biology must be changed, through a technological intervention that would begin with contraception and abortion but would end in the option of completely removing the reproductive process from women's bodies. *The Dialectic of Sex* proposes a post-revolutionary world in which human society has been transformed in order to deliver women, children and ultimately also men from the tyranny of an oppression that is rooted in biology. The nuclear family will have been disbanded. Human beings will live together in 'households' of seven to ten adults. Children need not be genetically related to any of these "parents" but will be raised by all, and will enjoy unprecedented freedoms. Machines will have rendered alienated labor obsolete, freeing people to pursue activities of their own choosing. Sexual difference itself will have been eliminated.

The radical feminism of Shulamith Firestone

This book argues for a return to Firestone. Today, nearly half a century after the *Dialectic*'s first publication, a resurgent feminism is again addressing the questions that Firestone raised. These are questions about the impact upon women of their role in procreation. But they are also questions about the arrangements in which we all cohabit, work, love and have sex. Who can be a mother? What does the nuclear family mean, for those both within and without one? What might be the effects of cybernetics on employment? What effects do ideas of romance and the erotic have upon our lives?

It is not surprising then, that recent years have witnessed calls to go back to Firestone's book. Nina Power and Laurie Penny, for example, have both urged today's feminists to turn to the 'deplorably overlooked' and 'criminally neglected' Firestone.[3] In 2010, Mandy Merck and Stella Sanford published an invaluable collection of academic essays on Firestone, *Further Adventures of The Dialectic of Sex*.[4] With the republication of *The Dialectic of Sex* by Verso in 2015, Firestone's text is now easily accessible to a new generation of readers.

To return to Firestone is, however, to reclaim a book that fellow feminist Ann Snitow has characterized as a 'demon text' of second wave feminism. *The Dialectic of Sex* has been constantly apologized for as exemplary of 1970s feminism's worst excesses and failings.[5] Subsequent feminists have criticized the book for *biologism*: for attributing to biology phenomena that it is thought are better understood as social or cultural in origins. It has been taken to task for *technological determinism*: for naively championing technological advance. Its assumption of the ubiquity of patriarchy has been called *dehistoricizing*. And critics have objected to what is taken to be Firestone's *abjection of the pregnant female body*: her construction of that body as an object of fear or repulsion. These and other criticisms will be considered in what follows.

And of course, there really is something difficult to swallow about its thesis. Something like Connie's affronted repudiation is to be heard whenever it is raised. Indeed, in the very first page of the *Dialectic*, Firestone anticipates 'the reaction of the common man, woman, and child – "That? Why you can't change that! You must be out of your mind!"' (3). For while there is much in the book that speaks the same language as twenty-first-century feminism, her radical proposals seem to have departed the field of rational debate. An end to the nuclear family? The abolition of wage labor? The creation of artificial wombs? Firestone's manifesto can seem both preposterous and hopelessly outdated:

a far-fetched, utopian hangover from a Swinging Sixties radicalism that has been definitively surpassed by the realism of subsequent decades. Firestone's revolutionary future can seem so fantastical that her book reads like science fiction.

What I am arguing for is therefore a *critical* return to Firestone. The flaws and the difficulties of the book are real. Some of the problems are to do with the peculiar history of the book and its author. Firestone was just 25 years of age when the *Dialectic* was published. She had become 'one of the most important architects'[6] of the radical feminist movement in northern America and the book was written 'in fervor, in a matter of months,'[7] while Firestone continued her grassroots organizing and activism. Upon publication it sold well, and was widely reviewed and discussed in journalism and on TV, where it was both 'lauded and excoriated.'[8] And yet it was never really 'taken up by grassroots radical feminism'[9] – probably as a result of its disagreements with much radical feminist thought, as we shall explore. Firestone disappeared from feminist politics and writing in the same year that the book came out, to concentrate on working as an artist. From then until her death in 2012, she seemed to eschew being identified with her earlier activism and writing (she was one of very few of the founding women's libbers, for example, who refused to be interviewed for Alice Echols's illuminating study of early US feminism[10]). One consequence of Firestone's withdrawal was that the *Dialectic* was thus in a sense orphaned, cast out upon the world to fend for itself without the guardianship of an author who would defend it against misunderstandings, correct or revise its arguments as feminist theory evolved over the years.

And of course, feminist theory has evolved, and in ways that reveal that the *Dialectic* has some major failings. For example, one of the achievements of black, lesbian and working class feminists has been to expose the ways that much second wave theorizing proceeded from assumptions based upon white, straight and middle class experience. The *Dialectic* all too frequently assumes that 'woman' is a unitary category, and in so doing installs just that white, straight, middle class experience as the norm. We will explore some of the problems this produces, in relation to Firestone's discussion of the family, and her deeply flawed and rightly criticized discussion of racism. But returning to her text might also cause us to ask about people whom she does not consider at all. For Firestone, a woman is someone with a womb. What then of trans women? Or of the pregnancies of trans men? Sometimes Firestone's text must be wrestled with – even read against the grain of her own intentions – in order that its insights be developed beyond the frame of the original analysis.

It is precisely the facilitation of this kind of critical engagement with the *Dialectic* that this book aims at producing. In so doing, I hope to help rescue what Mary O'Brien astutely calls this 'unevenly luminous' book[11] from its undeserved dismissal as the "mad one about artificial wombs." This attempt at rescue is inevitably partial. My focus is on Firestone's core thesis about reproduction, and this focus means that there are other aspects of her text – such as her discussions of romance and love, beauty and eroticism – that I have not been able thoroughly to address. I do hope, however, that readers will walk away feeling encouraged to also give *these* aspects of Firestone's work the critical attention they deserve.

Chapters 2 to 8 aim to return readers to the text itself, to what exactly is, and is not, being argued there. Doing so will require setting the thesis within the context of the urgent debates taking place within the radical political movements of the late 1960s and 1970s. And it will involve establishing the nature of Firestone's own critical engagements with a wide set of theorists that includes Marx, Engels, Beauvoir, and Freud. The aim of these chapters is primarily to present and to problematize Firestone's arguments, and to ask about the relationships between their different elements. I propose, for example, that as a methodological principle, Firestone's causal account of women's subordination be distinguished from her proposals for its solution. It is possible to hold that Firestone is right in the former, but wrong in the latter. But readers will see that I also want to offer a qualified defense of Firestone's thesis. In Chapters 9 to 10, I shall turn to what seem to me some of the most urgent issues of reproductive freedom facing us today, to offer a *Firestonian* account of how these might be thought about, and what might need to be done.

In defense of science fiction

This book, therefore, will offer a defense of science fiction, and especially of its utopian variant. When Firestone called explicitly for utopian science fiction and then sketched one herself in her (in)famous final chapter, she did so while understanding that utopianism can be dangerous. But she did so, nonetheless, since she understood as well the opposing but greater dangers of not believing that something better is possible at all. Utopian visions don't have to be about projecting a society deemed to be perfect, and still less about doing so with finality. But they matter – perhaps even are crucially needed – because they disrupt one of the key functions of ideology: that of making the status quo seem that it *could not be otherwise*.[12] Firestone's work, as good science fiction, speculates about the possible direction of a future society in order to show that our present one both could and should be better than it is.

Fundamentally, I shall argue that despite its blind spots and flaws Firestone's book is invaluable for feminist politics today. In a situation in which, as Nancy Fraser and others have observed, feminist discourse is so frequently appropriated into a neoliberal framework,[13] the *Dialectic*'s affront to "common sense" is precisely what is needed. Recoverable from among those second wave ideas that the march of hegemonic values has left behind are radical impulses capable of being reformulated in order to energize a genuinely oppositional feminism – one that cannot easily be co-opted into strategies for selling make-up, lingerie, sex toys and pole dancing lessons, or for justifying the waging of wars abroad.

The twenty-first century remains in thrall to ideologies of maternity and the nuclear family. We face a growing right-wing populism that continues to treat pregnant women as baby-vessels and to stigmatize non-reproductive lifestyles, even as the economic policies of neoliberalism make starting and supporting a family more difficult than ever for the majority of people. For anyone interested in the political significance of abortion, contraception, IVF, egg-freezing, surrogacy, changes to the nuclear family and to the nature of work today, there are few starting points as sharp, funny, rude, brave, clear-sighted and unapologetic as the work of Shulamith Firestone.

2

Rebellious Daughters of Chicago and New York

"Cool down little girl, we have more important things to do here than talk about women's problems."
(William Pepper)

These words, attributed to William Pepper at the National Conference for New Politics in 1967 (formed to unite leftist political organizations behind an anti-Vietnam presidential ticket), were delivered to the 22-year-old Shulamith Firestone as she and other women activists attempted to take the stage to propose a resolution on women's issues.[14] The resolution would have called for access to contraception and abortion; equal pay; the reform of marriage, divorce and property laws; and against the sexualization of women in the media. Pepper reputedly underscored his advice to Firestone with a kindly pat on the head.

Firestone was at this time living in Chicago, where she was forming connections with other politicized women. She was born in Ottawa, Canada, in 1945, to a German Jewish mother who had fled Nazi persecution, and to a father from an assimilated Jewish family, who had, however, embraced Orthodox Judaism in his youth. She was raised in the American Midwest and attended university in St Louis, Missouri, before moving to Chicago to study drawing and painting at its Art Institute. While there, she became the subject of a documentary on the 'Now Generation' by four male student filmmakers, in which she undergoes a 'brutal critique' of her work by her male lecturers. In her interview she declares her belief that 'any intellectual or artistic woman is "going to have problems with men."'[15]

The counter culture, radical politics and women

The political consciousness of the author of *The Dialectic of Sex* was forged in the late 1960s, in a period of manifest contradictions, many of which are articulated by Firestone herself. They have to do with the situation of women who, while being told that they had unprecedented freedoms and opportunities, were still living lives as inferior beings. They have to do with a perception, widespread among the younger generation, that there was a malaise at the heart of the Western world and particularly of America: that far from being the land of the free, America was a place of widespread economic injustice, of the systematic oppression of African Americans, of brutal military interventions overseas.

And they have to do as well with the situation of women within the counter cultural movements themselves. These often reproduced the structures of domination characteristic of the very mainstream culture that they protested against, expecting women to make the tea and to answer the phones but excluding them from decision-making. Issues of sexual equality were rarely on the agenda. Indeed, leftist radical movements were often actively hostile. In 1969, at the Counter-Inaugural conference in Washington DC, designed to protest Nixon's presidential inauguration, Firestone and another campaigner, Marilyn Webb, faced a struggle even to get women's liberation on to the program.[16] Webb's eventual speech was met with chanting by some male activists of '"Take her off the stage and fuck her!"' The response to Firestone was 'more feral' still, and rather than rebuking the hecklers the conference organizer tried to compel the women to leave the stage. Firestone later commented that '"a football crowd would have been…less blatantly hostile to women."'

After the 1967 National Conference, Firestone had gone on to co-found a number of women's groups, emerging as a driving force behind the burgeoning women's liberation scene. These groups admitted only women and they campaigned around the urgent issues facing women *as women*. The first was the Westside group in Chicago; and then, following Firestone's relocation to New York, she co-founded a number of women's groups there. New York Radical Women was the first women's group in New York City, and was established by Firestone and Pam Allen in 1967. In February 1969, however, frustrated that NYRW was not an explicitly radical feminist group and in the aftermath of the horrific experience of the Counter-Inaugural conference, Firestone formed the Redstockings with Ellen Willis.[17] The group's name was designed to recall the "bluestockings" of the eighteenth century (a pejorative name for intellectual and educated women) but adding to this tradition the '"red of revolution."'[18] In the fall of 1969, Firestone left the Redstockings to co-found, with Anne Koedt, the New York Radical Feminists, intended as an umbrella organization for the radical feminist groups that were cropping up across the state. Firestone is recalled by her fellow radicals as being fiercely intellectual and unafraid of confrontation. At a protest at the Ladies Home Journal, for example, where between 100 and 200 women activists occupied the offices and issued demands including the hiring of female and black personnel, the establishment of a day-care center, and the publication of an issue dedicated to women's liberation, Firestone met the reluctance of the male editor-in-chief by clambering onto his desk and 'shredding copies of the *Journal*.'[19] But her very outspokenness and intellectualism aroused resentment among some other members of the groups, who found Firestone elitist, dominating and dictatorial, and who objected to the limelight attracted by women who had written books within a movement supposed to be leaderless and egalitarian.[20] Firestone and Koedt finally suffered a rebellion against their Stanton-Anthony Brigade. All this no doubt contributed to Firestone's eventual withdrawal from movement politics.

Firestone's trajectory encapsulates that of many radical women in the period. At first a part of the loose coalition of hippies, students, anti-war activists, black civil rights campaigners, and anti-capitalists that made up the US counter culture, women eventually sickened of the marginal roles accorded them and the routine misogyny to which they were subjected, and left these movements to form their own ones. In so doing they pushed the emerging Women's Liberation Movement (WLM) in new directions.

The problem with no name

That movement had been kick-started in 1963 with the publication of Betty Friedan's bestseller, *The Feminine Mystique*.[21] Friedan had asked herself why so many women of her acquaintance who were supposedly living "The Dream" – hubby, children, suburban home, vacuum cleaner – in fact endured lives of quiet desperation. In the early 1960s women typically married very young and began families shortly afterwards. They were entering higher education in unprecedented numbers, but were studying in heavily gendered areas and were expected to abandon any careers upon getting married. Friedan's conclusion was that American women had been sold a lie. A life of stultifying routine and repetition, without outlets for creativity and intelligence, had been repackaged for them as the ultimate fulfillment of eternal womanly qualities. Her book became a publishing sensation, striking a chord with the thousands of women whom she had diagnosed as suffering from the 'Problem that Has No Name': an amorphous condition of dissatisfaction that – lacking any framework of political analysis – women interpreted as a problem or failing within themselves.

Friedan's book was a call for women to reject the housewife-as-destiny creed and to seek fulfilling lives outside the home, through paid employment and access to the professions. She argued that this required legal reform: rescinding laws that restricted access to contraception and abortion, but also campaigning for laws against the discrimination of women in the workplace and, especially, against unequal pay. In 1966 Friedan went on to found the National Organization of Women (NOW), which campaigned on these and other fronts to enlarge women's legal freedoms and to end discrimination against them in the workplace. This sort of feminism is usually characterized today as "liberal" feminism, for its reliance upon a tradition of liberal political thinking about the autonomy and rights of individuals.

What if capitalism itself is the problem?

By the late 1960s, however, the significance of liberal feminism was already waning. For many young women, the goals of NOW or similar campaigns were simply not far-reaching enough. Friedan's solution, for example, was that the middle class women she predominantly addressed could hire help to perform the domestic and child-raising tasks that they themselves eschewed. It is a "solution" that relies upon the continuing existence of a class of worker (probably female, often non-white) who performs alienated, poorly paid, and essential work. Friedan's feminism was about securing participation for women on equal terms with men. But what if the system itself were wrong?

Women on the political Left turned to traditions of socialist and Marxist thought to argue that the origins of the problems lay not in bad laws but in the economic basis of society. For these thinkers, the problem was capitalism itself: there could be no equality within capitalism because it is a system predicated upon the existence of an exploited class (whose surplus labor is expropriated by the capitalist as profit). A feminism that accepted capitalism could only be a movement seeking the enrichment of some women while the majority of other women (and men) remained oppressed. Socialist revolution was therefore the necessary condition of gaining true equality for women (and men). If previous socialist revolutions (in the USSR, for example) had not delivered gender equality, then this only showed the necessity of making women's issues central to the revolutionary campaign.

Radical feminism

Firestone's work is usually identified as belonging to radical feminism, a strand of feminism that, like socialist feminism, is "revolutionary" in seeking a wholesale transformation of society. Radical feminism, however, typically rejects the socialist feminist emphasis upon economics, and argues that the theoretical framework offered by classical Marxism is inadequate for understanding the nature and extent of women's oppression.

The 1970s radical feminists championed consciousness-raising groups in which women would come together in small groups without the presence of men, to discuss their experiences of the home, marriage, children, work and sex. By starting with the perceptions and feelings of women, it was believed, women would come to formulate their own tools for analyzing their situations, rather than having to rely upon the male-authored concepts of Marxist theory – which would inevitably determine in advance what aspects of female experience had political significance and in what ways. Radical feminists were also responsible for one of the key slogans of second wave feminism: "the personal is political." The formulation expresses the feminist insistence that politics is not something that occurs only in the so-called public sphere, stopping politely at the threshold of the home; but that instead, those aspects of our lives which we consider most intimate – whom we love, how we have sex, what we see and feel when we look in the mirror – are governed by versions of the relationships of power that structure society itself, and as such are both "public" and deeply political. Radical feminists argued that it was in fact one of the defining characteristics of their society that it was organized to ensure the subjugation of women to men, in the interests of men, introducing the term "patriarchy" to capture this idea. But if this was the true extent of the problem, they argued, then the liberal feminist idea of "discrimination" barely touched its surface. For radical feminists, women were not merely discriminated against: they were *oppressed*. Patriarchal culture had a distorting effect upon women, preventing the full expression of their potential, and transforming even their sexuality into something artificial that was shaped according to the dictates of male desire.

Despite Firestone's conventional labeling as a radical feminist she does in fact diverge in some important ways from other such thinkers, as comparison with the two other hugely influential books of 1970 – Germaine Greer's *The Female Eunuch* and Kate Millett's *Sexual Politics* – reveals. And importantly she was also (in her own particular way) a socialist feminist, yoking her proposals for changes in the sphere of sex and reproduction to a socialist demand for shared ownership of the means of production and for the distribution of resources according to need. One particularly important thing that Firestone's work thus reveals is the inevitable provisionality of the labels "liberal," "socialist" and "radical" feminist. These may be useful ways to begin to map the field of second wave feminist politics, but they quickly start to obscure as well as to illuminate its details.

Reformists versus revolutionaries

When Firestone herself came to map the strands of American feminism, in the second chapter of the *Dialectic*, she proposed that beneath this tripartite structure there lay a more important twofold distinction, between 'reformist' and 'revolutionary' feminisms. Rather provocatively, she placed the leftist/socialist feminists of her day (the 'Politicos') on the reformist side alongside liberal – or in her terms 'conservative' – feminists. Identifying the resurgent interest in feminist ideas as 'the second wave of the most important revolution in history,' (15) Firestone argued that since the nineteenth century the women's movement had always been divided between a genuine (because genuinely radical) feminism on the one side, and on the other a set of tendencies that channeled energies elsewhere and thus wasted their revolutionary potential.

Her key example is women's obtaining the vote. This had, she argued, initially been conceived as only a step towards obtaining political power, but it was seized by reformist women's rights campaigners as *the* priority, thereby marginalizing other issues of equal or greater importance. The eventual granting of the vote in 1920 in the US was, she argues, just an act of appeasement that conned women into accepting mere formal equality in place of substantive equality; and consequently dissipated the feminist struggle for the following 50 years. She describes the subsequent decades in terms of the permutations they offer of the 'Myth of Emancipation' – an ideology that works to 'anaesthetize women's political consciousness' (24) by affirming that, since they are now equal, if any woman should still feel anger, dissatisfaction or despair, it must be a personal failing and one that requires a private solution.

Unpleasant residue of the aborted revolution

Firestone writes searingly of the Sixties counter culture as one in which 'the boys,' fed up with the 'cloying romanticism' of the Fifties, have rejected monogamy and the traditional family but nonetheless want to have a 'chick' in tow – one who will never turn down sex (not be 'uptight') nor demand commitment (be 'a drag') (26–7). It is a faux radicalism, she contends, in which women are 'still invisible as people' (27). And it is one in which many people are channeling their anger into impotent forms of "politics" that ameliorate their personal frustrations while doing nothing to bring about real change. (This sounds a warning, perhaps, for today's Internet warriors.)

She claims that many of the 'rebellious daughters' of 1970 don't even know that there has been a feminist movement, so effective has been the 'blackout of feminist history' that distorted and misrepresented the movement, celebrating the reformers and willfully forgetting the radicals such as Susan B. Anthony, Elizabeth Cady Stanton and Sojourner Truth (28). But these daughters are rebellious nonetheless, since they are confronted with the contradictions that are the 'unpleasant residue of the aborted revolution': they have near full legal freedoms, but no power; educational opportunities, but are unable to use them; supposed sexual freedom, but one that is an intensified sexual exploitation, enacted through the circulation of their bodies between (supposedly radical) men, and the circulation within the wider media of endless 'hateful or erotic images of themselves' (28–9). For Firestone the feminists who identify with various forms of leftist politics (the 'Politicos') do not offer any genuine solution to these contradictions. Unwilling to insist on the centrality of women's issues, they instead try to fit them into the 'existing leftist analysis and framework of priorities – in which, of course, Ladies never go first' (33). In the *Dialectic* Firestone would do the reverse, fitting a socialist analysis into a feminist analysis that prioritized issues of sex and reproduction. For her, any putative feminism (conservative or leftist) that fails to start with these issues will achieve superficial reforms at best, since it will have failed to tackle the roots of women's oppression.

So what were the issues around sex and reproductive control that faced women when Firestone was writing in 1969?

Crimes against chastity

At one point in the *Dialectic*, Firestone notes acerbically that contraception is not the same thing as 'family planning' (185). While widely offered to married couples who could be understood as seeking merely to delay starting a family (following a 1965 Supreme Court ruling), in 1969 birth control was legally available to *un*married women only in some states in the US, reflecting a moral insistence that the purpose of sex was procreation. Until 1970 it was possible even for references to contraception to fall foul of federal anti-obscenity laws. For example, a campaigner, Bill Baird, was jailed for three months under Massachusetts' "Crimes Against Chastity, Decency, Morality and Good Order" law after he handed contraception to a female Boston University student, Sue Katz, as a political action at a talk he'd given on birth control and abortion. In some respects Baird was lucky, since under this law he could have received as much as ten years in prison. Suitably unchastened, he continued to fight and in 1972 won the Eisenstadt v. Baird Supreme Court judgment that legalized contraception for all Americans.

Finding Jane

In 1969 abortion was illegal in all but a handful of US states, and these had decriminalized it only in relation to very specific circumstances, such as where pregnancy resulted from rape or incest, or where continuing to term would threaten permanent disability or death to the mother. In consequence, illegal, unsafe, so-called "back-street" abortions were rife; and were disproportionately sought by women on low incomes, who could not afford in-patient hospital stays and were unlikely to know personally doctors who would help them.[22] The numbers of women who died through these back-street procedures are highly disputed, but NARAL: Pro-Choice America estimate it was as many as 5000 each year.[23] One such woman was Geraldine Santoro, in 1964. Santoro had fled an abusive marriage and on becoming pregnant by her married lover had feared that her husband would kill her and her children. Failing to obtain an abortion any other way, Santoro's lover eventually performed the procedure on a motel room floor with borrowed tools and instructions, before abandoning Santoro when it became clear that things had gone wrong. Her corpse, slumped forward over bloodied towels, was discovered the following morning by a chambermaid. The police photograph of the scene testifies to a death as horrific as it was unnecessary. After it was published in the feminist *MS* magazine in 1973, the image was mobilized as a national pro-choice symbol. Santoro was 28 years old, and left behind two daughters, who had been told that their mother had died in a car accident.

By the late 1960s women's liberation groups were responding to this situation with direct action. In February 1969, feminists interrupted the proceedings of New York State's legislative hearings on abortion reform. As Echols reports, a panel of 15 experts had been invited to give evidence, consisting of 'fourteen men and one woman – a nun.'[24] After one of the experts had proposed allowing abortions to women 'who had "done their social duty" by having four children,' one of the protestors stood up and demanded to hear from '"some *real* experts – the women."'[25] Also in 1969, a group of women in Chicago formed "Jane," the pseudonym for their Abortion Counseling Service. Jane members started out by helping pregnant women to find an abortionist and assisted during the procedure. Eventually learning that the abortionists, who were always men and always charged an extortionate price, were usually not doctors at all, the women themselves started performing low cost abortions, the payment to be decided by what the woman determined she could afford.[26] Jane's phone number was distributed in movement leaflets, on notice boards, by sympathetic friends or even professionals, and women would call "her" to leave their contact details and be phoned back. The Jane operation was busted by police in 1973 and seven members were charged with battery, although the case was eventually dropped in the wake of the Roe v. Wade judgment that established a legal right to abortion in the first two trimesters of pregnancy.

But while Firestone was writing *The Dialectic of Sex*, a sexually active woman stood in significant danger of unwanted pregnancy and of prosecution, injury or death in dealing with it. Compulsory motherhood was a very real prospect indeed.

A manifesto

The Dialectic of Sex was above all else a manifesto – a public declaration of intent. It is a book that declares that a feminist revolution *must* happen, and that it *can happen now*, since for the first time in human history the technology exists to enable the root cause of women's oppression to be addressed just as the contradictions of women's situation have emerged with such painful clarity. And like all manifestos it is characterized by 'compression' and 'hyperbole.'[27] It does not mince its words to make either itself or its author agreeable to anyone whom its analysis may offend. It rips into the Sixties romanticization of nature, for example, by describing the cult of natural childbirth as 'only one more part of the reactionary hippie-Rousseauean Return-to-Nature, and just as self-conscious.' It infamously describes childbirth as being 'like shitting a pumpkin' (181). It is a book that announces that anyone is wasting her time who thinks that an apologetic or deferential feminism is going to change anything.

So why did Firestone argue as she did; and is she right? It is to these two questions that the next few chapters turn.

3

Women's Oppression Is Natural

Anyone observing animals mating, reproducing, and caring for their young will have a hard time accepting the "cultural relativity" line.
(Firestone, p.9)

Firestone's daring move is to tackle head on a subject that feminists before her had often shied away from. For many other thinkers, to concede that women's oppression could be natural in origin would be to give up the fight as lost. The reasons for this are perhaps obvious. Defenses of male dominance are almost always couched in the language of what is "natural." For countless male philosophers of the Western tradition, women were "naturally" less capable than men of the kind of abstract rational thought required for participation in political decision-making, and as such could be justly excluded from it. Anti-feminists (both male and female) in the nineteenth century opposed the extension of suffrage to women on such grounds and argued that women's natures suited them to the domestic and not the public sphere. In the twentieth century, second wave feminist demands to equalize the roles of the sexes in the home and workplace were met with the objections that women were the natural homemakers and carers of children and that men were the natural breadwinners. Today, we often hear that the overwhelming predominance of men in politics, business and in the higher echelons of almost all public organizations reflects natural differences between the sexes and the natural disposition of men to be more assertive and risk-taking than women.

To allow "natural" differences to be the frame for thinking about the oppression of women therefore seemed to many feminists to be conceding the terms of the enemy. It is not surprising then, that so many of them eschewed nature as a possible source of women's subordination, preferring instead to point to law (liberal feminists), economics (socialist feminists) or culture (other radical feminists) as the fundamental problem.

The "natural" is not a human value

Firestone's distinctiveness is to recognize that what is natural is thereby neither necessarily good nor inevitable. She is implicitly drawing a distinction between explanation and justification (and this, as Stella Sandford also notes, is all too often overlooked in discussions of Firestone[28]). She agrees that women's oppression is *explained* by natural conditions or differences. But she disagrees that these natural conditions *justify* that oppression. She does this in virtue of her claim that 'the "natural" is not necessarily a "human" value' (10). This is a concept central to Marxism, but also to the existentialist philosophy of Simone de Beauvoir, and it is Beauvoir whom she quotes for a fuller elaboration: '"Human society is an antiphysis – in a sense it is against nature; it does not passively submit to the presence of nature but rather takes over the control of nature on its own behalf"' (10).

What Beauvoir and Firestone are saying is that if "nature" is taken to mean something like "the physical environment as it exists without human intervention" then it is very clear that human beings do not for the most part consider nature to be an intrinsic good. It may be natural to die of exposure in a thunderstorm, or when one suffers severe blood loss from an injury, or when the cells of the body divide in cancer, but we rarely consider these things *good*. Nor do we consider them unchangeable. Instead we set about building shelters, making bandages and performing blood transfusions, and the manipulation of chemicals and radiation. Nothing, it seems, is *less* human than to let nature take its course.

When it is claimed that inequalities between men and women

are natural, and when such claims are used to oppose attempts to achieve equality, then however "nature" is being used not only to *show* how that inequality has come about (explanation), but also to *justify* it. It is to say that it should be allowed to continue without corrective efforts because what is natural either cannot be changed or ought not to be. And this is a view that no one really holds who has ever put up an umbrella in the rain.

Let's change nature

Firestone's response to the litany of claims that women's oppression is natural could be characterized as "Yes it is, so let's change nature." In my view, it is a response that alone would justify her status as a foundational feminist thinker. Earlier feminist attempts to evade the question of nature, to say that there are no salient natural differences between men and women, or at least that we cannot know whether there are until men and women are treated in the same way, were always a hostage to fortune (however mistakenly). What if science were to prove that there really were natural differences and that these played a role in producing inequality? Firestone's point is that this simply doesn't matter, since human beings have the power (perhaps uniquely among animals) to intervene purposively in nature and to change it. On her view, people need to decide what kind of society is desirable and then act to bring that into being – "with" *or* "against" nature.

Her recognition that the "natural" is without intrinsic value underpins the arguments of *The Dialectic of Sex* and explains her affirmation of the liberatory potential of *technology* – understood here as the process of intervening in nature in the name of human values and goals. Firestone might have made a slightly different argument against "nature," namely that most cases of usage of the term are actually incoherent. As philosopher John Stuart Mill argued in his essay 'On Nature' from 1874, either "nature" means that which occurs spontaneously and in the absence of human agency, in which case there is very little that is natural; or "nature" means everything that unfolds according to the laws of nature, in which case everything that happens is natural, including all human action, and nothing could be unnatural, including all forms of human intervention in the physical world. The enormous value of Mill's essay lies in his systematic unpacking of the utter absurdity of holding that one ought to do what is "natural," and ought not to do what is "unnatural." As he points out (along with Firestone, Beauvoir, and Marx), what is "natural" on the first definition is in most cases held by human beings to be contrary to our goals, and so the unnatural is precisely what we ought to do. And what is "natural" on the second definition makes doing something unnatural literally impossible.

Firestone does not analyze the incoherence of the "nature" term in quite this way. Her usage seems to accord with the first of Mill's senses that I have given above: human reproduction is 'natural' to the extent that it remains unaffected by forms of human intervention such as contraception (which are 'technology'). It is in this sense that her feminism is pro-technology and anti-nature. She nevertheless mounts a remorseless assault on the functioning of "natural" and "unnatural" as terms that gain their force and meaning from the ideological systems that they support rather than from real differences in the world. To give just one example, she describes how participants in a Harris poll had proved surprisingly open to considering new reproductive technologies such as test-tube fertilization when employed in the service of the traditional family, 'to help a barren woman have her husband's child,' but rejected the proposal that they could be used to extend reproduction beyond the nuclear family (179). As Firestone notes, 'it was not the "test tube" baby itself that was thought unnatural...but the new value system, based on elimination of male supremacy and the family' (180). This is just as true today, when techniques such as surrogacy or egg transplantation are used to extend reproductive capacities to gay couples or to older women and are denounced as being "unnatural." The term tells us nothing about the procedure or technology itself, but much about the moral disapproval felt by the speaker.

But why does Firestone believe that women's oppression is natural? The answer has to do with her acceptance of the radical feminist claim that male domination is transhistorical and transcultural: in other words, that all human societies that have ever existed have been patriarchal. Firestone argues that for this to be the case it must be that the cause lies in something deeper and more fundamental than law, economics, or cultural phenomena, since these can be expected to have varied across time and place. And what it is that is deeper and more fundamental, and shared across all human history as well as the animal kingdom, is the division of females and males into animals that do, and animals that do not, bear offspring.

4

Sex Class Is the First Class

Unlike economic class, sex class sprang directly from a biological reality: men and women were created different, and not equal. (Firestone, p.8)

The core thesis is this: throughout most of human history women have been at the 'continual mercy of their biology' (9). Before the advent of reliable contraception or abortion, the post-pubescent female could expect to spend perhaps 30 years of her life (should she live so long) pregnant, giving birth or nursing small children, as well as suffering the 'female ills' associated with her reproductive system (menstruation, menopause, etc.). These facts, together with the protracted period of helplessness of human infants and their consequent need for care and supervision, entailed a severe curtailment of women's capacity to take part in the productive labor that produced food, resources and wealth. Because children were dependent upon women, women were therefore dependent upon men for provision of the things needed for physical survival. This division of labor is driven by 'natural reproductive difference[s],' but it quickly becomes reinforced by human action (9). Men, Firestone claims, seek to fortify the power they are granted through women's dependence and then to extend that domination wherever possible to other men. Men and women have been divided into two distinct classes by biology (producers and reproducers), but this fact in turn produces a psychological formation – a desire or need for power – that leads to the incessant formation of further divisions of humanity into unequal classes, castes or 'races.' It is in this sense, then, that for Firestone the oppression of women is natural: it is rooted in a reproductive biology that, for millennia, it has not been within the power of human beings to control.

However, this also means that, while in an important sense women's oppression is natural, it is also not *simply* natural, since it has always involved the fortification of a natural inequality through the decisions that men have made to consolidate their power. We can see this, for example, in the case of childrearing. Pregnancy and childbirth may be biologically given, and so too may be the nurturing of infants, to the extent that until recently natural breast milk was the only food source for babies. But the delegation to women of caring responsibilities for older children is a cultural artifact, a non-biological extension of their "natural" reproductive roles.

Indeed, more importantly still, I think, one can read Firestone as arguing that since technological development has now reached a stage where it is possible to intervene in and change human reproduction, even the contribution to women's oppression made by those biological facts can no longer be characterized as "natural." It may be clearest to demonstrate this by analogy. For as long as human beings lack the technological knowledge and/or resources to prevent, for example, a flood, then the destruction of a village by flooding may be described as a "natural disaster." In a situation in which the expertise exists, however, to build dams and other flood defenses, but this has not been done, or not done sufficiently well, then the flooding of the village can no longer be characterized in this way since the disaster has come about as a result of human decisions, actions and inactions. If "nature" means for Firestone something like "that which lies outside of the scope of human control," then it may be true that for millennia human reproduction and thus the sexual inequality she thinks it entails have been natural. But in the late twentieth century, with medical science having advanced to the point that it has, her claim is that reproduction and hence the inequality it spontaneously produces do now lie within our control. If that is so, then, just like a village flooding when the technology exists to prevent it, the continuation of sexual inequality is not natural – it is cultural and political, since it is the result of human decisions, actions and inactions.

Why dialectic?

Why is a book that puts forward this thesis entitled *The Dialectic of Sex*? The answer lies in Firestone's relationship to the dialectical or historical materialism of classical Marxism. Among the feminists of the 1960s who had turned away from the liberal tradition, judging its supposed solutions too superficial, were many who had turned instead to Marxism. In the theory of historical materialism developed by Marx and Engels in the nineteenth century, the movement of human history could be explained as the product of class struggle. The young Karl Marx had been heavily influenced by the philosophy of G. W. F. Hegel, who saw world history as evolving through a process whereby a social or cultural formation, proving flawed in some way, generated an opposing formation which nonetheless preserved aspects of the former one, before itself yielding to a new formation which again arose in contradiction. But where Marx had agreed with Hegel that history progressed through a dynamic interplay of opposing forces (the dialectic), he rejected Hegel's 'idealist' conception of this as consisting in the coming to self-realization of some kind of universal rationality (Hegel's 'Geist'). Instead, Marx had insisted on a materialist understanding of the dialectical movement of history, arguing that the forces concerned were economic, having to do with how human beings produced the physical resources they needed in order to live. Marx's materialist philosophy held – in contrast to that of Hegel and many others – that it is not primarily ideas that dictate the concrete forms in which human beings live; rather, it is these material conditions which determine the ideas and values that people hold (*being* precedes *consciousness*). He believed that material conditions under advanced capitalism had reached a point where the exploited majority (the proletariat) could achieve consciousness of themselves as an oppressed class in whose interests it was to overthrow their rulers and seize direct control of the means of production.

Quite whether Firestone's theory *is* dialectical has been a subject of debate by commentators, as has the issue of whether her idea of sexual *class* works, in Marxist terms.[29] What is clear, however, is that she calls it dialectical in order to signal both her ambition to produce a materialist account of history, and her sense of a schism that is inaugurated by biology and then played out in myriad ways throughout human culture.

Marxism and second wave feminism

Marx's theory held out hope to many of the feminists of the 1960s who had come to believe that the oppression of women ran so deep that nothing short of revolution could remove it. But his insistence on the principal role of economics provoked more ambivalent responses. On the one hand, feminist thinkers already alert to the multiple ways in which women were economically exploited in capitalist society found that aspects of Marx's work offered powerful tools for illuminating this. His concept of a reserve army of labor, for example, could express the way that women formed a pool of surplus laborers capable of being brought into industry when male labor was scarce (for example in times of war) and then easily expelled back into the home when it was not. Feminists also noted how the labor of women assisted capitalist production in another way, which had not been fully recognized by Marx: that their unpaid reproductive labor in the home both provided the capitalist with the next generation of workers in the form of children, and maintained the current adult labor force by providing male workers with the food, clothing and emotional care that they required in order to labor efficiently each day in factory or office. But on the other hand, many feminists – even some who identified as

Marxist or socialist feminists – felt that classical Marxism was too hamstrung by its economic framework; too reductive in its insistence that oppression came down in the final instance to economic factors; and too blinded, therefore, to the facets of a woman's oppression that could not be revealed through the lens of economics.

Firestone's response to classical Marxism is typical of the audacity and irreverence of her book. Agreeing with other radical feminists that Marx's is a reductive theoretical framework that it would be damaging for feminism to 'squeeze' itself into, she does not however reject it (7). Instead she aims to incorporate it into her own feminist analysis and hence to correct it – to correct its dialectic.

Engels on the Origin of the Family

Her starting point is the work of Friedrich Engels. Engels had offered a more systematic analysis of women's oppression than had Marx, in an essay that was being taken up in the 1960s by socialist feminists. *The Origin of the Family, Private Property and the State* (1884) was written by Engels after Marx's death, to systematize Marx's notes on the work of American anthropologist, Lewis Henry Morgan. Morgan's *Ancient Society* (1877) was a treatise on the evolution of human society through three epochs of savagery, barbarism and civilization – thought to correspond broadly with changes in the form taken by the human family. For Morgan, the nuclear family first appeared in the final years of barbarism, becoming fully established only in the phase of civilization.

The great value of this work for Engels was that it historicized the nuclear family, suggesting that it was not an absolute and unchanging fact of human existence, but only a contingent one. Even more than this, Morgan believed that transformations in human living arrangements were driven by economic factors – or in Engels/Marx's terminology, changes in the modes of production. The different epochs which corresponded to different family structures were themselves characterized by successive ways of appropriating resources from nature. Morgan had, Engels believed, arrived independently at a materialist understanding of history that supported Marx's own.

Morgan's anthropological work thus gave sustenance to Engel's project of discursively destabilizing the nuclear family. As early as *The Communist Manifesto* (1848), Engels and Marx had identified the bourgeois nineteenth-century family as a viper's nest of hypocrisy and unfreedom: an economic unit designed to manage the inheritance of property, that was shrouded in a veil of sentimentality. They considered that the unnaturalness of monogamy led almost inevitably to male adultery, while bourgeois wives had effectively to prostitute themselves to husbands, the sale of their sexual services in marriage being their only way of making a living. In showing that the nuclear family was but a contingent historical fact, developed out of economic conditions that themselves would prove impermanent, Engels was therefore able to show, he believed, that the family itself would disappear under conditions of communism, to be replaced by unions voluntarily undertaken and based on love.

Engels' materialist analysis of the family also enabled him, he believed, to show that the oppression of women had a distinct historical origin. He was here writing explicitly against the pronouncements of fellow Marxist thinkers August Bebel and Karl Kautsky, who held the oppression of women to be something that had existed 'from the beginning of time' (Bebel).[30] Drawing on Morgan, Engels proposed that the earliest human communities had been characterized by a sexual freedom enjoyed by both women and men. Indeed, he believed that women had been endowed with a high degree of social authority, since the only reliable way of establishing a child's lineage was through the mother. This was all to change, however, with the emergence of private property.

Engels conjectured that these early human communities had been communistic, collectively owning those few goods that were not immediately consumed, and that had been hunted or foraged from the land, or later, produced through primitive agriculture. As the modes of production developed, however, more goods were produced that were surplus to subsistence requirements, and these became privatized as the belongings of particular individuals. The very shift to heavy agriculture that enabled this increase in production, also made it more difficult or dangerous for women to participate in production (which they had done previously to a very full extent), particularly if they were pregnant or accompanied by small children. Since production now had the capacity for generating wealth, it became valorized over reproduction; just at the same time that women were increasingly relegated to the latter.

Engels believed that with private wealth now a reality, and conscious of their newly acquired superior status as producers, men sought to replace matrilineal with patrilineal inheritance. But the only way to do this would be if paternity could be known for certain. They therefore sought to eliminate women's promiscuity by binding them within a strict monogamy (while continuing to enjoy sexual freedom themselves). As such, women became privatized as being among the possessions of a particular man, and their children and reproductive labor with them. The patriarchal nuclear family was born, and it represented what Engels emphatically called 'the world historical defeat of the female sex.'[31]

Firestone on Engels

Socialist feminists in the 1960s had considered Engels' account of an original human sexual equality giving way only with the emergence of private property to be both compelling and hopeful, since it identified not only the cause of women's oppression but also a solution. They echoed Engels' call for women to achieve economic independence from men by re-entering the sphere of productive labor, and argued for feminism as a necessary complement to Marxism, since both theories recognized the overthrow of private property as the necessary condition of human emancipation.

For Firestone, however, this was a *necessary* but not a *sufficient* condition: it was required, but on its own it would not be enough. For her, Engels had not inquired sufficiently deeply into the material realities of how human beings sustain life. 'There is a whole sexual substratum of the historical dialectic that Engels at times dimly perceives,' she writes, 'but because he can see sexuality only through an economic filter, reducing everything to that, he is unable to evaluate it in its own right' (6). We might wonder if this is quite fair, since, as we have seen, Engels does make sexual reproduction a key part of his analysis. But Firestone's point is that this is so only at the point that there is surplus production and therefore wealth to be inherited. Prior to that, questions of pregnancy, birth and childrearing do not appear to him to have been salient. Indeed, he does not consider them to have mattered for gender equality, claiming that while women have always taken predominant responsibility for children, under primitive communism this division of labor did not entail *inequality* (reproduction being no less valued than production during this economic phase). It is this view that Firestone finds untenable. For her, the different reproductive functions of women and men have always resulted in a division of labor that, since it entails a power imbalance, therefore generates domination. Thus, even in the primitive stages of human history, women constituted an exploited class. If Engels cannot see this it is because he 'acknowledged the sexual class system only where it overlapped and illuminated his economic construct' (7).

Like other radical feminists, then, Firestone sees Engels' understanding as fundamentally limited by his adoption of economics as the lens through which to view all aspects of life. She therefore rejects his supposition of an historical origin to gender inequality, holding that instead, since reproductive biology has always been the way it is, women have always been oppressed.

Ahistorical Firestone?

This is the major claim that has led to Firestone being criticized for being *dehistoricizing* in her analysis. Firestone, so the argument goes, takes features of her own society – the nuclear family, patriarchal relationships – and projects these back upon the past. In so doing, she fails to appreciate how human relationships are historically varying: how, for example, women's experiences of pregnancy and childbirth are different under different social arrangements.[32]

Is this a problem for Firestone? This is something that the reader should consider. To aid that reflection, however, there are two aspects of Firestone's theory that must be clarified.

The first is that in her own terms Firestone's work is absolutely not ahistorical. Rather, she aims to show that the social unit within which reproduction takes place *has* changed; that these changes have gone hand-in-hand with changes in production; and that it is reproduction and production together that have been the drivers of history. The biological family is not the nuclear family. The former is something like the biological unit of male sperm-giver and female ovary/womb-provider and genetic offspring: *this* unit has remained unchanged throughout history (until, potentially, now). The latter, however, is just one form that the social institutionalization of this biological unit might take: married, (supposedly) monogamous parents with the father as head of the household and children in a position of legal minority. And this form of the biological family is historically fairly recent.

The second necessary clarification is that Firestone is,

however, committed to the view that beneath this historical variation certain 'contingencies' (10) have always been at work: that women have always been rendered less powerful than men through their role in reproduction. She is therefore not claiming that all societies have been male-dominated *in exactly the same way*. But she is claiming that all societies of which we have knowledge have evidenced *some form* of male domination (she includes matriarchies in this). And she is claiming that the reason this is so is that the biological family itself is characterized by power imbalance, which imbalance is then transmitted in some form or another to whatever social form the reproductive unit then takes. The nuclear family does, however, have a certain privilege in Firestone's analysis, in that it particularly *reveals* – by *intensifying* – the inequality of the biological family itself.[33]

Indeed, Firestone appropriates Marx in her attempt to outline an historical framework for understanding reproduction and the social injustice it leads to. She praises Marx for seeing how history evolves as a dialectical process, in which forces to do with how human beings maintain life in the face of particular environmental conditions, act and react upon one another. This materialist emphasis gives his account a scientific status that, she believes, is missing from earlier pre-Marxist socialisms and from the feminisms of her time. She aims, however, to offer an analysis that is *more materialistic* than Marxism. Where Marx and Engels see culture as being driven by economics (production), Firestone posits a material determinant beneath even this, and that is biology itself (reproduction). The penultimate chapter of her book offers a two-page diagram in which she maps out the progress of human history, showing how cultural forms and achievements have correlated with particular phases of production, and these in turn with different forms of social organization of the biological unit of the family. It is a schema that stretches, however, beyond a history of the past and the present, and into the periods of 'revolution,' 'transition' and 'ultimate goal' that she projects for the future (172–3).

Revolution

This revolution too is a modification of the Marxist schema. Where Marx and Engels posited that the proletariat must realize their oppressed condition and revolt against the ruling class by taking control of the means of production, Firestone amends this in her own distinctive terms. For her, any revolution that is a revolution of production alone and not also of reproduction will have failed to uproot the ultimate source of the problem. In common with many other radical feminists, she argued that the socialist revolutions of the twentieth century had failed because they did not 'eliminate the family and sexual repression' (190). Any revolution that leaves intact the basic structure of the family will see 'any initial liberation' quickly 'revert back to repression' (190), for the biological family is the 'tapeworm of exploitation' (12).

Instead, a revolution of production must go hand-in-hand with a revolution in reproduction, which would consist of women taking direct control of the means of reproduction through reproductive technologies. Just as Marx believed that only a particular historical juncture would make possible a successful communist revolution (industry must have progressed to the point where nature can be made to yield the resources needed for all), so Firestone believes that a feminist revolution can happen only now, at the stage of technological development witnessed by the advanced capitalist societies of modernity. These have delivered first of all reliable and safe contraception and abortion, and are on the cusp, she thinks, of a technological innovation that could liberate women altogether from the burden of the biological reproduction of the species.

TFTFY

Firestone's first chapter finishes with a *literal* rewriting of Engels.

She has quoted earlier his definition of historical materialism from his book, *Socialism: Utopian or Scientific*:

> Historical materialism is that view of the course of history which seeks the *ultimate* cause and the great moving power of all historical events in the economic development of society, in the changes of the modes of production and exchange, in the consequent division of society into distinct classes, and in the struggles of these classes against one another. (5)

She now represents this 'strictly economic definition' (5) in a form corrected in light of her own thesis that it is biology and not economics that is the material stratum of history:

> Historical materialism is that view of the course of history which seeks the ultimate cause and the great moving power of all historic events in the dialectic of sex: the division of society into two distinct biological classes for procreative reproduction, and the struggles of these classes with one another; in the changes in the modes of marriage, reproduction and child care created by these struggles; in the connected development of other physically-differentiated classes [castes]; and in the first division of labor based on sex which developed into the [economical-cultural] class system. (Words in square brackets are in Firestone's original.) (12)

It is a tactic of quoting and editing that is now familiar in the age of Internet forums – TFTFY: There Fixed That For You. Firestone has now laid out the full scope of her ambition, which is for a feminist incorporation of Marxism that, rather than rejecting its conclusions, will see these as being true within a limited field of application, rather as (she somewhat grandiosely says) 'the physics of relativity did not invalidate Newtonian physics so much as it drew a circle around it' (7). This is therefore also an incorporation of socialist feminism within radical feminism.

Firestone's biologism?

What is quite distinctive about Firestone is that she supplies a materialist explanation for the apparent ubiquity of male domination. In so doing, she demystifies it. She explains it in a manner that in no way justifies it or sees it as immutable. She says that male domination is a result of how women's ability to take part in wider social life has, historically, been severely curtailed by their role in reproduction. And she says that this is something that can be changed.

Firestone is therefore claiming that there is a biological component to something that other feminists have taken to be purely cultural. For this, she has been criticized for biologism, or biological determinism. Donna Haraway, for example, has charged Firestone with making 'the basic mistake of reducing social relations to natural objects'; which mistake – Haraway thinks – then leads Firestone into a dangerously reckless championing of technological control over nature (see Chapter 7 for more on this).[34] Michelle Barrett has also worried that Firestone's account falls into 'biologistic assumptions,' wondering whether '"feminist biologism"' can escape the problems of other biologisms – such as suggesting that there is little hope for change.[35]

But for Firestone, precisely the point is that without attributing biology some causal role, the *ubiquity* of male domination remains unexplained. And, I would add, because it is unexplained it precisely is therefore *mystified*. There is no accounting for why it should be the case that *all* societies (or even, if one wants to argue for exceptions, *most* societies) are, and have been, male-dominated. And precisely because there is no explanation, this phenomenon becomes available to other explanations that do seek to claim the correctness and immutability of male rule: to propose, for example, that it is the consequence of an innate male superiority.

Interrogating Firestone

My question for the reader is whether Firestone's explanation of the emergence of male domination is plausible. To count as plausible, it does not need to be capable of proof beyond doubt. It does need, however, to account for all relevant observable facts, to offer conclusions that follow logically from its premises and to be without internal contradiction.

What might be weaknesses in her argument? We have seen that it relies upon particular steps in its etiology of oppression. The first is the claim that bearing children makes women dependent upon men, to the extent that men are required to supply resources such as food and shelter. I think Firestone should be interpreted as making a fundamentally historical argument here: that this was the case, in the past, when fertile women were in an almost constant cycle of pregnancy, birth and nursing, and in which this inevitably limited their capacity to participate in production on equal terms with men. Is this a reasonable supposition? Are there grounds on which it might be contested? The second step is the claim that this dependency then leads to the development of particular psychological formations: men, experiencing the dependence of women on them, seek to reinforce the power that this gives them, and to extend it to other men as well. For me, this is perhaps the more vulnerable of the two argumentative moves. Why might men not have responded to women's reliance on them with tenderness or compassion? Why might they not have recognized their own reciprocal dependence upon women, for the bearing and raising of their children? Is Firestone here relying upon an implicit claim that there is only one kind of likely response to the dependency of women and children: that having power, men get used to having power? How secure is this supposition? And is it a claim about how each and every individual man might respond? Or is it a claim about the kind of culture – the set of general assumptions, attitudes and values – that might develop under such conditions?

As I proposed in the Introduction, one must separate out Firestone's explanation (of the cause of women's oppression) from her proposed solution. Even if is she is right that pregnancy and childbirth lie at the origins of women's oppression, is she therefore right that removing this oppression requires a transformation in biology?

To better understand Firestone's objections to pregnancy and childbirth, we will turn now to her engagement with Simone de Beauvoir.

5

Like Shitting a Pumpkin

Let me then say it bluntly: Pregnancy is barbaric.
(Firestone, p.180)

These words occur in the final chapter, where Firestone is discussing the need to 'free humanity from the tyranny of its biology' (175). She is writing in frustration at what she sees as the Left's failure to perceive the urgency of developing reproductive technologies that will not only liberate women from the bondage of traditional reproduction but will also enable the species to avert ecological disaster through over-population (a prominent concern in the late 1960s, and one that is woven throughout Firestone's text). She notes approvingly the existence of artificial insemination and inovulation, comments acerbically that male as well as female oral contraception might be a reality were not male scientists such worshippers of male fertility, and looks forward to test-tube fertilization, sex selection, and the development of artificial placentas that could support a fetus through to viability. In advancing this program of *controlled, artificial,* reproduction, she understands that she is confronting a powerful set of cultural fears and taboos. Not least of the resistance will come, she knows, from within the WLM itself, where 'many women...are afraid to express any interest in [reproductive technologies] for fear of confirming the suspicion that they are "unnatural"' (180).

We therefore return to one of the text's dominant themes: that the "natural" is not a human value. For Firestone, that "nature" is frequently barbaric and something that human beings would do well to escape is never more evident than in the case of pregnancy and childbirth. Much of her objection has to do with the facts of physical suffering and pain. '[C]hildbirth hurts,' she writes, 'and it isn't good for you' (181). If *The Dialectic of Sex* is mostly remembered today for one other thing besides its proposal for artificial wombs, it is its (in)famous description of childbirth as being 'Like shitting a pumpkin' (181). Firestone tells us that this was related to her by a friend who is a mother. She then imagines a conversation continuing between this friend and what she calls the School of the Great-Experience-You're-Missing: 'What's-wrong-with-a-little-pain-as-long-it-doesn't-kill-you?' retorts the School, obviously not envisaging that any satisfactory insistence on there being something wrong is possible (181). But the problem, as Firestone will insist, is that it is not a matter of just a little pain, and it not infrequently does kill you. In the distant past, she writes, childbirth involved both agony and risk to life; and this was explicitly recognized – and women were 'admired in a limited way' for enduring it (181). But in modern societies a 'mystification of childbirth' has taken hold, according to which its physical consequences are shrouded in sentiment, and women are obliged to adopt 'proper' attitudes – 'as in, "I didn't scream once"' (181).

Firestone is here taking aim at a dominant tendency of the counter culture to romanticize "nature" and an earlier, supposedly more primitive, simpler and more harmonious way of life. 'Natural childbirth,' she writes, 'is only one more part of the reactionary hippie-Rousseauean Return-to-Nature, and just as self-conscious' (181). In 1969, the same year that Firestone wrote these words, British novelist Angela Carter would publish her post-apocalyptic novel, *Heroes and Villains*, in which a renegade professor (modeled on Timothy Leary) reigns over a tribe of 'barbarians' whom he has fashioned after his reading of Enlightenment philosopher Jean-Jacques Rousseau. Firestone's point seems to be much like Carter's – that ideas about the innocence of man in his natural state are dangerous fantasies; that the Nature that is to be Returned To is a misleading intellectual fabrication drawn up according to the demands of a philosophical or ideological system; and that the desire to replace complexity with simplicity tends towards authoritarianism.

But Firestone is also writing against a tendency within the WLM itself, and one that will become more dominant as the 1970s progress, to celebrate the facts of sexual difference, and in particular, the biological processes of pregnancy and childbirth (on which, more shortly). She states that she cannot accept the view of many of her fellow feminists that pregnancy is beautiful; that, if it is not widely seen as such, this is only because of the distorting lens that patriarchal values bring to our perceptions. In a somewhat curious passage (curious not least for its apparent valorization of *natural* reactions over learned responses) she writes:

> The child's first response, "What's wrong with that Fat Lady?"; the husband's guilty waning of sexual desire; the woman's tears in front of the mirror at eight months – are all gut reactions, not to be dismissed as cultural habits. (180)

Evidently, the issue here is not simply the pain and danger of childbirth, but the suggestion that an intuitive and almost *aesthetic* aversion to the pregnant body shared by men, women and children alike is revealing of something fundamentally important. Things are clarified in the next sentence, which is key to Firestone's thesis in several ways: 'Pregnancy is the temporary deformation of the body of the individual for the sake of the species' (180).

The influence of Simone de Beauvoir

Firestone is here drawing heavily upon Simone de Beauvoir (to whom the *Dialectic* is dedicated) and her account of female embodiment in *The Second Sex* (1949). Beauvoir's book is a landmark text of the feminist tradition, and Firestone was not alone among second wave feminists in crediting it as a major source of inspiration. Beauvoir argues that women in contemporary society are secondary beings – creatures who are in actuality inferior to men, who are defined through their relationships with men (fathers, lovers, husbands) and who allow men to establish for them their roles in life. But she also argues that this condition – *femininity* – is neither essential nor innate: it is not the "nature of women." 'One is not born, but rather becomes, a woman,' wrote Beauvoir in her famous formulation, meaning that anatomical females are induced to assume femininity through a process of social conditioning that starts from infancy.[36] And crucially for Beauvoir, this becoming-woman is not *determined*. Beauvoir was bringing to bear upon her analysis of the situation of women the existentialist philosophical framework of her friend and intellectual collaborator, Jean-Paul Sartre. According to this, whatever the concrete circumstances into which human beings are born, they retain a freedom to think, feel and act according to goals and values that they establish for themselves. She thus argued that whatever the pressures exerted upon women to assume femininity (inferiority) these could not have the force of *necessity*: women retained the capacity to resist such conditioning and to assert themselves as full human beings on an equal footing with men. *The Second Sex* was a rallying call for women to do just that.

Much of Beauvoir's account of women's biological being occurs in an early part of the (800 page) book, where she is trying to show that neither biological, economic nor psychological factors can causally determine women's social condition. It is therefore designed to show that nothing about a woman's physical makeup makes it inevitable that she should be subordinated to men. It is a discussion that was to prove deeply controversial, and its bearing on Firestone's ideas makes it worth exploring in some detail.

The purport of Beauvoir's account is that their reproductive functions make women the 'victim of the species' (52). The basic idea is that there exists a conflict between the needs of the particular individual and the needs of the species to which they belong. Evolution has it that the species needs the individual to procreate; but this may come at a cost to the individual themselves, either in terms of physical health, or, in the case of human beings, in terms of interfering with their life goals. The individual cost of species perpetuation is not, however, borne equally by men and women. Beauvoir's judgment is stark: women are sacrificed in this conflict by their biologies; a woman's reproductive system afflicts her as a series of crises which work against her individual interests at almost every stage of life.

She proceeds to prove this by cataloguing the physical and emotional ailments associated with menarche, menstruation, conception, pregnancy, childbirth, lactation and menopause. Periods are 'a burden, and a useless one from the point of view of the individual' (60). Of fertilization, she writes, 'First violated, the female is then alienated – she becomes, in part, another than herself' (54). She consistently depicts the growing embryo as a parasitic and alien presence that drains the woman of vital nutrients and strength ('gestation is a fatiguing task of no individual benefit to the woman') and understands miscarriage and morning sickness as more or less successful rebellions against 'the invading species' (62). She emphasizes the 'serious accidents or at least dangerous disorders [that frequently] mark the course of pregnancy'; on childbirth, she observes that 'the infant...in being born...may kill its mother or leave her with a chronic ailment'; and describes breastfeeding as a 'tiring service' in which the 'mother feeds the newborn from the resources of her own vitality' (62–3). Beauvoir's point throughout this discussion is to establish that in enduring the trials of pregnancy, birth and breastfeeding a woman's body almost ceases to be her own: it is hijacked by the species for the assertion of its own claims.

The words she uses most frequently to describe a woman's role in reproduction are *violation* and *alienation* – expressing on the one hand a sense of the rupturing of bodily integrity, and on the other the estrangement or expropriation of one's body from oneself.

When Firestone writes that *pregnancy is the temporary deformation of the body of the individual for the sake of the species* she is surely channeling Beauvoir. To deform is to make misshapen. Neither Firestone nor Beauvoir can see the pregnant body as just one form that a woman's body may take, and they certainly cannot see it in a positive light. Instead, it has about it the character of distortion and aberration: its meaning is that the woman has fallen victim to the species.

But there are also crucial differences between Beauvoir and Firestone. Beauvoir never goes so far as to propose that reproduction could or should be removed from women's bodies. One might suggest that a reason for this is the different historical moments in which the two texts are written: Firestone's was produced during a period of accelerated development in reproductive technologies, which perhaps allowed such a radical solution to become conceivable in a way that it hadn't been 20 years previously. But there are theoretical differences that also have a bearing.

Beauvoir, as I said earlier, holds that none of these facts about a woman's embodiment *determine* her to become inferior. They are not the *cause* of her oppressed condition. This might seem a surprising conclusion, so rhetorically powerful is the litany of female ills that she provides. But Beauvoir is able, or perhaps compelled, to argue this because of the existentialist position she adopts in relation to the facts of biology. Biological facts, for her, do not have intrinsic value. That is to say that the meaning of any particular fact of biology is not contained within the fact itself, but is dependent upon the context within which the fact appears – and that context is irrevocably at least in part the outcome of human decisions and goals. For example, it may be a physiological fact that women on average have less muscular strength than men, but in a society where 'violence is contrary to custom' (67) this will have no significance and will not be to women's disadvantage. She makes exactly the same argument in relation to women's reproductive biology, claiming that 'The bearing of maternity upon the individual life…is not definitely prescribed in woman – society alone is the arbiter…[and] individual "possibilities" depend upon the economic and social situation' (67). Societies can choose to lessen the burden of reproduction on women by demanding fewer births and by providing better medical care during and after pregnancy and childbirth.

Firestone would of course agree that this should be so, but for her such measures will almost certainly be insufficient. For Firestone, contra Beauvoir, the value really does reside within the fact. For her, the fact that it is the woman's body and not the man's that must bear the burden of reproduction *does* deterministically set up a power hierarchy and as such it is the fact itself that must be altered. In the *Dialectic*'s opening chapter she warmly praises Beauvoir for being 'the only one who came close to…the definitive analysis' of women's oppressed condition and for recognizing that economics could not be the explanation (7). But claiming that 'sex class sprang directly from a biological reality,' she goes on to write that:

> Although, as De Beauvoir points out, this difference [biological difference between men and women] of itself did not necessitate the development of a class system – the domination of one group by another – the reproductive *functions* of these differences did. The biological family is an inherently unequal power distribution. (8)

For Firestone, then, men's and women's different reproductive functions, rooted in biology, *necessitate* domination, and the biological family is *inherently* unequal. Biology is causal; it is therefore the explanation. She will not accept Beauvoir's own ultimate explanation of female oppression, which is that women have elected to accept their Othering by men because it procures certain compensations for them: the economic compensation of having their living provided, and more importantly the existential one of having men determine for them their goals and values (a flight from the absolute freedom that on an existentialist account is commonly experienced as anxiety and burden). Indeed, Firestone views the Hegelian concept of Otherness on which Beauvoir's explanation depends as just one more obfuscating philosophical category whose emergence must itself be explained through a materialist analysis that starts with sex and reproduction, and she considers *The Second Sex*'s reliance on its existentialist framework to be the book's weakness (as have many subsequent feminist commentators). For Firestone, since it really is biological reproduction that is the cause of women's oppression, it is biological reproduction that must be changed.

Abjecting the pregnant body?

Neither Firestone nor Beauvoir can see anything positive about a woman's role in reproduction. Although Beauvoir does not agree that women's reproductive role is ultimately causative of their oppression, she does admit it to be a serious limiting factor in a woman's life and the best she can say of it is that its deleterious effects can be ameliorated through social and economic choices. But is either thinker justified in construing pregnancy and childbirth in such deeply negative terms?

A charge frequently made against both Beauvoir and Firestone is that they are viewing pregnancy and childbirth from a patriarchal perspective. Indeed, a persistent criticism that has been made of *The Second Sex* is that many of its arguments proceed from an identification with masculine values and culture. Something that is frequently advanced as evidence of this is Beauvoir's view of mothering (the raising, rather than the bearing of children). She says explicitly that mothering is not a *project* – the existentialist term for those human actions that are undertaken with the conscious design of bringing about substantial change, of opening up a new future. It is instead about stagnation and repetition – the maintenance of the species – and as such is mere animal activity. It lacks the creative import of those acts by which the existential hero transcends the given conditions of his existence (among which Beauvoir controversially includes acts of killing and war). Here and elsewhere Beauvoir – or so critics have argued – values activities associated with men, and sees little value in those behaviors or qualities that have been traditionally the preserve of women. Her vision of equality is one in which women would become more like men. It is a charge that has ironically led to the author of one of the founding texts of feminism being accused of misogyny.

Much of Beauvoir's characterization of female biology is certainly very contestable. She is keen to establish, for example, that the existential meaning of sexual intercourse is different for men and women. Men, she thinks, discover a sense of their power over the world through penetration and thus intercourse is for them an affirmation of their individuality and autonomy. For women, however, penetration (even when desired) means submission and violation: 'In this penetration her inwardness is violated, she is like an enclosure that is broken into...Her body becomes, therefore, a resistance to be broken through' (53). Although such descriptions occur in a chapter entitled 'The Data of Biology,' suggesting objective, neutral description, they are disturbingly reminiscent of certain fantasies that structure male-authored pornography. And she seems in this discussion to have forgotten her own existentialist insistence that the value of a biological fact is not intrinsic to it: for Beauvoir, female embodiment hinders women from achieving their goals.

Different social circumstances might enhance or diminish this effect but there is no circumstance in which the meaning of their biology is not *limitation*.

Is Beauvoir guilty then of *abjecting* the pregnant body? Is Firestone? The charge here is that both writers discursively construct this body as an object of fear and repulsion, when in reality it is no such thing. Are their perceptions being skewed by the values of a patriarchal culture that construes women's differences from men – especially bodily ones – only in negative terms; even while they set themselves against patriarchy?

Against Firestone and Beauvoir – feminists celebrate maternity

Firestone's views on maternity had never been shared by the majority of radical feminist activists, and from 1970 a different way of thinking about women's reproductive role gained prominence. According to this, the pregnant body was beautiful, and women's reproductive biology was to be celebrated as a source of distinctively feminine values and power. The activist Jane Alpert, for example, in 1974 published her manifesto *Mother Right: A New Feminist Theory*.[37] In this she praises Firestone for her account of sex oppression as being foundational of all oppressions. But she writes that Firestone's analysis of biology is 'deficient,' since it fails to recognize that 'female biology is the basis of women's powers.' For Alpert, the capacity to bear children – the *very consciousness* that one's body is capable of creating new life, even if a woman never becomes a biological mother – is the source of distinctively female psychological qualities, such as 'empathy, intuitiveness, adaptability... protective feelings towards others and a capacity to respond emotionally as well as rationally.' Biology, for Alpert, is 'the source and not the enemy of feminist revolution.'

In the 1980s, radical feminism developed into "cultural feminism," which sought to celebrate the ways in which women differed psychologically from men. Many cultural feminists argued, as Alpert had done, that women's biological differences played a determining role in producing these psychological differences, leading women to experience and make sense of the world differently. One claim was that women had a kind of natural affinity for progressive politics. For example, the menstrual cycle that Beauvoir considers a useless monthly trial was reconceptualized as keeping women in tune with the seasons of the natural world, therefore inclining women towards environmentalism. Where Beauvoir saw pregnancy as alienating and the fetus as an invader, many cultural feminists held that gestating life allowed women to understand the mutual dependencies that constitute all being, therefore disposing them to cooperative ways of living and to pacifism.

In the meantime in France, a new kind of feminism (to which Beauvoir would object) had emerged that also celebrated sexual difference and "femininity," drawing upon psychoanalytic theory and French poststructuralism. Hélène Cixous argued that for over 3000 years Western patriarchal culture had systematically denigrated femininity, associating it with negativity and death. She called for women to rescue their bodies from these associations, and to give them new significations through a practice of creative writing that she called 'writing the body' or *écriture féminine*. Cixous's own densely allusive and poetic writing enacts this practice, and continually works through images that figure the female sexual and maternal body in terms of creativity and generosity. For Cixous, maternity permits women an experience of *giving* (milk, caresses, love) in which one gives freely and for the sake of the other: something normally disallowed in patriarchal capitalism, where one gives only in order to obtain a return. Of pregnancy and childbirth she urges: 'Rather than depriving woman of a fascinating time in the life of her body just to guard against procreation's being recuperated, let's de-mater-paternalize.'[38] Cixous suggests that in reclaiming their bodies from the patriarchal imaginary, women stand to recuperate a special relationship to literary creativity that has its origins in the capacity of a woman's body to create new life: 'How could the woman, who has experienced the not-me within me, not have a particular relationship to the written,' she asks.

Biological determinism and somatophobia

To the extent that they all seek to affirm the ways in which women are thought to differ – emotionally, psychologically and physically – from men, all the above positions can be broadly classified as *gynocentric* feminisms.

Some of them are also biologically determinist – in that they allege a simple correspondence between anatomy and psychology – and in ways that I think Firestone definitely is not. For Firestone, biological sex *in itself* does not have any immediate consequences for psychology. She is not someone for whom, for example, aggression is hard-wired into men through their higher levels of testosterone. She does not claim that the character of male genitalia makes men psychologically disposed to rape. It is actually sex class inequality that, for her, leads to the development of aggression in men, and to subservience in women, and she is very clear (as we shall see) that once sexual inequality has been eliminated, anatomical differences between the sexes will cease to be significant. She would have regarded the kind of position advanced by Alpert, I think, as what she was to call in *Airless Spaces* 'matriarchalist theory…a glorification of women as they are in their oppressed state.'[39]

These gynocentric positions are complex and varied, and are not all biologically determinist (some see women's psychological differences as arising from their socialization). But they are linked by a call to reassess women's embodiment outside of a framework established by patriarchal culture. From this perspective, both Firestone and Beauvoir can be criticized for failing to do this.

Philosopher Elizabeth Spelman sees Firestone's text as evincing what she calls *somatophobia,* or fear of or disdain for the body.[40] For Spelman, somatophobia is a key element of the way that oppressive discourses such as misogyny and racism operate, where the supposed physical differences of the Other are singled out for particular disdain or repudiation. Faced with a long history of patriarchal thinking that negatively associates women with their bodies, placing men on the side of "culture" and women on the side of "nature," feminists have a number of ways to respond, Spelman says. One way is to demand for women's embodiment to be rethought and resignified outside of the incessantly negative terms of patriarchal ideology. This would be what the cultural feminists and Cixous are doing. For Spelman, both Firestone and Beauvoir take a different route, and it is a deeply dangerous one. It is to view 'embodiment as a liability' and to seek, therefore, to break the identification of women with their bodies. The problem is that while this sets itself in opposition to the patriarchal devaluing of women, it nonetheless accepts that tradition's devaluing of the body itself. Beauvoir and Firestone are examples of feminists who 'think and write as if we are not embodied, or as if we would be better off if we were not embodied.' And in assuming and endorsing that somatophobia, she argues, they fail to challenge a key element of not just misogyny but also other, overlapping, oppressions including racism. 'Flesh-loathing is part of the well-entrenched beliefs, habits, and practices epitomized in the treatment of pregnancy as a disease,' Spelman writes. 'But we need not experience our flesh, our body, as loathsome.'

Reassessing Firestone

I think that the question of whether Firestone is right in her characterization of pregnancy and childbirth as *barbaric* is logically separate from the question of whether her causal explanation of women's oppression is correct. This is partly because Firestone's causal account works (if it does work) at a different scale from that of personal experience; but it is also because the causal account makes a claim about *function*, where the normative account makes one about *value*. It is entirely possible for it to be true that at the level of the human species, women's role in reproduction has been crucial, historically, in establishing societies based upon male domination; *and* that at the level of the individual, particular women have experienced their pregnancies not as barbaric but perhaps even as creative, fulfilling or empowering. Firestone's argument does not require that every woman is made dependent through pregnancy and childbirth, or that every man respond to female dependence with the desire to reinforce domination. It only requires that in the past, female dependence through reproduction, and a consequent predominance of men over other social functions, has been sufficiently widespread to produce economic, legal and cultural arrangements that further entrench gender inequality – and that continue to do so even after the originating contribution of biology has been diminished in significance.

Firestone would not, perhaps, have fully agreed with me here. She seems to think that there is an objective meaning to female reproductive biology and that it is negative. For me, however, the very fact that Cixous and others can just as persuasively describe the physical processes of maternity so differently shows that how these are *thought* (what significance and value they are made to bear) varies individually and culturally. There is perhaps no "right" answer – in the sense of objectively true – to the question of what *value* female reproductive biology has, or what the experiences of pregnancy and childbirth *are like*. But, to repeat: this is different from the question of what function that biology has had, historically, at the level of the species.

And surely, feminist challenges to the authority of purely negative characterizations of women's reproductive biology –

that say, "there are other ways of thinking that biology" – ought to be welcomed?

Does this mean, however, that Firestone's characterization of pregnancy as barbaric is patriarchal? That she is guilty of somatophobia? Of flesh-loathing? I'm not sure there are simple answers to these questions. On the one hand, it is certainly true that Firestone sees women's embodiment as a problem, perhaps *the* problem. But on the other, she may actually be less vulnerable to the problem of somatophobia than are others, such as Beauvoir. Although Beauvoir's negative characterization of female embodiment extends to pretty much all aspects of a woman's physical being, including muscular strength and lung size, Firestone's comments are restricted to the processes involved in pregnancy and childbirth, and usually have to do with the suffering caused to women therein.

That these processes *do* involve suffering is undeniable. A woman going through pregnancy and childbirth is likely to endure at least some of the following: morning sickness; exhaustion; hemorrhoids; agonizing labor pains; damage to pelvic muscles, nerves and ligaments; and perineal tears. She risks injuries that might lead to ongoing or even lifelong pain, fissures, bladder or bowel incontinence and diminished sexual enjoyment. Death in childbirth remains a substantial risk in many parts of the world, and even with the healthcare provision of technologically advanced Western societies it is far from zero (indeed, in the US it is actually increasing[41]). It is estimated that in 2015 alone, 30,300 women died as a result of pregnancy or childbirth worldwide.[42] Firestone reminds us that this suffering and these risks are only the outcomes of a contingent natural process (that evolution has developed *this* way and not another). It has not been ordained according to some providential wisdom that women should so suffer in the course of reproducing the species (except in the view of a very few Christian fundamentalists). It is not a *good thing*.

In offering this reminder, she opens up an important discursive space. It is a space that allows women honestly to discuss the physical aspects of maternity and the fears they may have in relation to them. All too often, such a space is closed down, and with it, the ability of women to make informed decisions about pregnancy and about pre- and antenatal options. In the UK and US, the medical professions seem strongly to favor vaginal over C-section births, and arguably to coerce women into the former. A "natural childbirth" ideology puts women in competition with one another to endure labor without pain relief. A wider culture of woman-blaming pervades discussion of maternity, operating to stigmatize or to shame any woman who thinks that she may be more than a baby incubator. Women who desire, or need, technological interventions, such as C-sections and pain relief, too often face censure, and are labeled trivial, vain, not brave enough, phobic, career-obsessed, privileged ("too posh to push"), unfeminine and unmaternal.

We remain in the grip of a cultural mythology that, as it had been in Firestone's day, shrouds maternity in romanticizing, sanitizing representations. Firestone's discussion of pregnancy may be one-sided, intemperate, and even possibly offensive, but it at least punctures the sentimentalizing veil to say that childbearing is often difficult and sometimes even traumatic. And it says that this matters, because women matter, in themselves, and not just as baby carriers. Her unfashionably negative discussion of pregnancy and childbirth reminds us that "natural" does not equal "good," that "technological" does not equal "bad"; and that the implications of this are that the threats to women's health and well-being entailed by replenishing the species should not be complacently accepted but call instead for urgent consideration, action and resourcing.

6

Against the Nuclear Family

This then is the oppressive climate in which the normal child grows up.
(Firestone, p.44)

Parental love misfires. This is one of the key themes of Firestone's book. Damaged parents produce damaged offspring, who go on to replicate the harm when they become parents themselves. Firestone seeks to explore just how it is that, in her view, the institution of the family produces this intergenerational trauma, and to propose a radical alternative for the rearing of children. In so doing she was to be instrumental in developing an analysis that became widespread in radical feminist circles in the 1970s and beyond, and that identifies the family as one of the chief sites and sources of a woman's – and child's – oppression.

Reading Firestone, reading Freud

To understand Firestone's critique of the family we must begin where she does – with the psychoanalytic theories of Sigmund Freud. Firestone proposes that Freudianism is perhaps the 'cultural current' that has most influenced life in America in the twentieth century, and that nobody can remain unaffected by its language of "neurosis," "penis envy" and "repression" (38). But she advances a feminist rereading of these Freudian categories. It is a rereading that purports to show that the processes of psychological development that Freud described are not to be understood as the immutable consequences of a timeless conflict between civilization and the individual psyche and its drives, but as the specific effects of a particular form of social arrangement – that of the patriarchal nuclear family.

To bring Freud and feminism together at all was in 1970 a fairly unusual maneuver. For many feminists, Freud was the enemy. There are ample reasons for this. Freud famously claims that women unconsciously desire a penis and resent their felt anatomical inferiority; he theorizes that their establishment of a superego is weaker than is the case in men, and hence that their capacity for abstract ethical judgment is diminished; he proposes that the meaning of motherhood is that a woman desires a child as the unconscious substitute for her father's penis. Generally in his theoretical apparatus, women feature as the weak link in humanity's achievement of culture, always threatening to pull the male subject back into the realm of nature and incest. The male child is almost always the starting point of his descriptions of human psychological development, and when the mature Freud comes to realize that the girl's trajectory cannot simply be the mirror image of her brother's, he responds in seeming irritation, attributing to female sexuality an inscrutability that he does not pause to consider might be the result of the gendered bias of his own analytical framework.

Why then does Firestone think there is something of value in Freudianism, and that feminism would be ill-advised simply to reject it as a patriarchal discourse? The answer is that she sees Freudianism and feminism as having a common origin in their recognition of the fundamental importance to human life of sex. '*Freudianism is so charged,*' she writes, '*so impossible to repudiate because Freud grasped the crucial problem of modern life: sexuality*' (40). '*Freudianism and feminism grew from the same soil*' of oppressive Victorianism, 'characterized by its family-centeredness, and thus its exaggerated sexual oppression and repression,' and both are reactions against this (41). Both are '*made of the same stuff,*' (41) she writes, meaning that both grasp the experiences that befall the sexuate being as the material basis of human individual and cultural development. Indeed, the title of Firestone's chapter describes Freudianism as 'The Misguided

Feminism.' Freudianism starts with the same insight as does feminism into the terrible damage wrought upon human beings by a family structure predicated on the (sexual) domination of most of its members. But then it goes wrong, loses its way, since it fails to realize that the solution is not a therapy aimed at the adjustment of the individual to this structure, but the abandonment of the structure itself.

It is the family that lies at the root of the problem, for Firestone, since this is fundamentally an organization of power and subservience that demands the channeling of sexual feeling into highly restricted paths. As we have seen from earlier chapters, for Firestone the family cannot help but have such a character because of the inequality that springs from the biological reality at its very basis. But since this biological reality can be changed, so too can the family that is causative of such widespread individual and cultural sickness be dismantled. In failing to recognize this, Firestone believes, Freud mistakenly sees the conflict between civilization and individual happiness as an eternal feature of human life, and thus his therapy seeks to reconcile the individual to this reality.

In fact, argues Firestone, the very reason for the widespread popularity of Freudian psychoanalysis in the latter half of the twentieth century is thereby disclosed: Freudianism as a clinical practice is imported to the US in order 'to stem the flow of feminism' (61). The aborted revolution of first wave feminism produced a generation of women stuck halfway between their traditional feminine roles and an alternative that had not yet been realized: the resulting frustration and confusion of these women 'often took hysterical forms' and 'sent them in droves to the psychoanalysts,' who in turn sent them 'scurrying back "adjusted" to their traditional roles as wives and mothers' (62). The meaning of not just psychoanalytic practice, but of all psychological therapeutic intervention in this period is, according to Firestone, that of inducing individuals to accept their circumstances rather than continue railing against them. As such, psychology becomes a conservative, even counter-revolutionary, force, and not just for women but for men and children as well. Quoting radical theorist Herbert Marcuse, to whose own analysis she is greatly indebted, Firestone describes therapy as having become '"a course in resignation"' (59).

This discrimination of political value between the *theory* and the *therapy* of psychoanalysis has a long history in radical cultural theory, and it attempts to show that while the latter is a bourgeois individualistic practice, the former has a revolutionary potential, since it points to the existence of some enclave within the self (the unconscious, the drives) that has not been colonized by the logic of patriarchal capitalism and that threatens to turn rebellious against its oppressor. Whatever one thinks of this proposition (and the psychoanalytic categories it employs) there is surely something very relevant today about the critique of psychological therapy, or at least the particular form that it often takes. In both the US and the UK today, someone suffering from psychological distress is more likely to receive cognitive behavioral than psychoanalytic therapy, but Firestone's charge perhaps applies just as well. Such therapies focus upon the individual, seeing the locus of the problem as being within them, in their difficulty adjusting to particular realities. Whether there might be realities to which one precisely ought not to adjust – of racial discrimination, for example, or poverty – is not a question that falls within the purview of the analysis or practice. Individuals are asked to change how they think and feel about the world, rather than to think about whether or how that world ought to change. In British higher education there is currently a vogue for training in "resilience," which means helping students and staff to cope with their stress. No matter how well-intentioned the trainers, it is a "solution" to an increasingly widespread psychological distress (anxiety and depression) that displaces attention from likely structural causes

– student debt, graduate unemployment, and proliferating workloads and precarious employment in higher education – and onto the individuals themselves and their capacity to endure the intolerable.

The family and power

In Firestone's reinterpretation of Freudian categories, it is power dynamics that are decisive. She writes that it is only through the filter of feminism that Freudianism makes sense. The Oedipus and Electra complexes and "penis envy" shed their logical inconsistencies and manifest absurdities only when considered from the point of view of the operation of domination within the patriarchal family. I will argue, however, that the absurdities of which she accuses Freud are in large part a consequence of her rather incautious reading of him, and that her feminist 'restatement' of psychoanalysis (56) actually amounts to a junking of its fundamental concepts. But I shall also argue that this does not matter, for what is really of value in Firestone's discussion of Freud is not the "Freud" part at all, but what she has to say about the family.

Let's take the Oedipus complex. Firestone introduces this as 'the cornerstone of Freudian theory, in which the male child is said to want to possess his mother sexually and to kill his father' (43). She grants that a boy (Freud means children under six) will desire his mother, but objects:

> it is absurd what Freud's literalism can lead to. The child does not actively dream of penetrating his mother. Chances are he cannot yet even imagine how one would go about such an act. Nor is he physically developed enough to have a need for orgasmic release. (46)

The problem here, however, is not so much Freud's literalism as Firestone's. Freud does not mean that the small boy knows what penetrative sex is or has the desire to accomplish this; he means that the boy behaves as a jealous lover. Because the mother is his primary care-giver, she is the object of his most intense longing, and he desires to have the mother's affection entirely to himself, to have at his constant disposal her attention and her physical caresses and to share these with no one else. Nor does Freud make the mistake of thinking that the infant wishes to "kill" his father, in the sense of the word that an adult would understand. This intense love of the mother makes the father appear in the light of a rival, whether because he intervenes to separate the child from his mother or because his mother also shows him (the father) her affection. Freud writes starkly in *The Interpretation of Dreams* that the child who formulates a wish for his father to be dead has no conception of corpses in the ground or not existing forever. In his imperfect understanding, typically gleaned from a grandparent's decease, to be "dead" means only to be gone, and the child is certainly capable in his passionate jealousy of forming *this* wish in relation to his father, even though it is likely to coexist with an attitude of love towards his father and a paradoxical desire for his presence and affection too.

But Firestone insists that the Oedipus complex makes sense only when read *metaphorically*, by which she means in terms of power. The boy desires his mother, but more importantly he also identifies with her, since she is like him in her powerlessness. In the prototypical nuclear family, says Firestone, the husband/father is the breadwinner, and:

> all other members of this family are thus his dependents. He agrees to support a wife in return for her services: housekeeping, sex, and reproduction. The children whom she bears for him are even more dependent. They are legally the property of the father...whose duty it is to feed them and educate them, to 'mold' them to take their place in whatever class of society to which he belongs. (44)

She is actually here describing the family of the late-nineteenth century, but subsequent sentences establish that the nuclear family in its more recent versions is not fundamentally changed, consisting of 'essentially the same triangle of dependencies' (44). Firestone acknowledges that the modern wife, unlike her Victorian forebears, may be highly educated and in employment, but claims that her dependency on her husband remains, reinforced by legal and economic arrangements, but driven ultimately (as always) by her role in reproduction:

> she is rarely able, given the inequality of the job market, to make as much money as her husband (and woe betide the marriage in which she does). But even if she could, later, when she bears children and takes care of infants, she is once again totally incapacitated. (44)

The child born into such a structure 'is sensitive to the hierarchy of power' (44). She/he understands that they are completely dependent upon the two parents, but they also see that it is the father who is in 'total control,' and that the mother resides 'half-way between authority and helplessness' (45). The boy child sees 'in most cases' that the father bullies his mother and makes her unhappy, and as such he bonds with his mother in their shared oppression. But he also comes to understand that this identification binds him to that oppression. Since he is precisely a *male* child, his father, whom at first the boy is likely to fear as a distant and authoritarian figure, will come to offer him a place in the privileged world of the male – what Firestone calls, quoting Erich Fromm, 'the exciting world of "travel and adventure."' 'Most children aren't fools,' writes Firestone. 'They don't plan to be stuck with the lousy limited lives of women' (47). So the boy has to '"repress"' (Firestone's quotation marks) his attachment to his mother and his hostility to his father: he has to 'abandon and betray his mother and join ranks with her oppressor' (47).

It is a betrayal that will leave him guilt-ridden, and that will distort his relationships to women in adulthood, to whom he will relate through the 'good/bad women syndrome, with which whole cultures are diseased' (54). Good women resemble the mother, and as such cannot be the objects of sexual feelings. Those women who do elicit such feelings must therefore be bad. 'A good portion of our language degrades women to the level where it is permissible to have sexual feelings for them,' she notes (54).

Firestone thus rewrites the Freudian Oedipal drama in terms of the dynamics of power that operate within the bourgeois family unit. She does a similar thing when it comes to the sexual development of the girl. The Electra Complex (in fact, Freud hardly uses this term) occurs, according to Firestone, when the little girl, who also initially loves her mother more than her father, transfers her identification to her father and 'rejects her mother as dull and familiar' (48). She does so because of her awareness that her father has access to a wider world of experience than does her mother, and just like the boy, she desires this world for herself. If a girl should see a penis (perhaps her brother's) and wish that she had one, this is not because there is anything enviable about this protuberance of flesh in itself. Rather, as an obvious physical difference between herself and the brother whom she has come to perceive is valued more highly than herself, and given greater freedoms, it takes on the power of explaining this differential treatment. Firestone here anticipates a great deal of feminist psychoanalytic work which, refusing simply to dismiss "penis envy" as laughable male fantasy, instead interprets it in social rather than biological or purely psychological terms. The penis is envied not for itself, but for the privilege that in a patriarchal culture it seems to the child to symbolize or even obscurely explain.

Nothing of what is said here about the shaping of a child's mind within a structure of gender-based domination seems to me implausible. But nor does it seem particularly Freudian. I agree here with psychoanalytic feminist, Juliet Mitchell, and with Rosemary Delmar (editor of the 1979 edition of the *Dialectic*), both of whom argue that Firestone actually sidelines the key ideas of psychoanalysis, such as *the unconscious* and the strength of irrationality.[43] Freud had theorized that rationality took only the smaller share in the functioning of our minds. To a much larger degree our attitudes and actions are motivated by ideas and wishes that are hidden to us, that perhaps have their origins in an early infancy that has been distanced from us not only through the passing of time but more significantly through a repression that sinks "unacceptable" recollections into a sea of amnesia, but that survive and have influence through a chain of unconscious associations, images and memories. As such, for him, much of what we feel and do is precisely *irrational*, because the forces that induce us so to think or act lie beyond the reach of our consciousness and therefore our capacity rationally to sort or assess them. But there is not much sense of this in the 'Freudianism' that emerges after Firestone's 'restatement' of it. In fact, as Mitchell and Delmar also note, the choices made by the little boy and girl of Firestone's re-descriptions appear as largely conscious and eminently rational responses to their circumstances. It is not that psychoanalytic concepts have been reformulated, or alternatively interrogated and refuted, but that they have been in large part eschewed in favor of an analysis that could proceed perfectly well without them.

The insecure, aggressive/defensive, obnoxious little person we call a child

For Firestone, the family, as the institutionalization of a gendered imbalance of power, must inevitably produce relationships that harm those within them. But this is in many ways not really the fault of mothers or even fathers, since the psychological damage they themselves accrued as children means they really cannot help but return the favor. Boy children, asked to turn away in contempt from their mothers and sisters, become adult men who must maintain this precarious superiority through assertions of ego, and who invest in their own children as securing 'that continuation of name and property which is often confused with immortality' (44). Mothers, meanwhile, having their access to the world beyond the home severely curtailed, make their children the center of their lives, placing upon them an intolerable burden, especially as the child grows and desires for itself a sphere of freedom outside of parental/maternal control. Laurie Penny has noted the reluctance of most feminists to acknowledge 'the real harm done by women as well as men in the domestic sphere,' as mothers 'have handed down suffering, guilt and the expectation of patriarchal servitude to their children.'[44] But Firestone is a clear-sighted exception here. She insists upon how, for as long as women are valued in a patriarchal culture primarily for their role in raising men's heirs, mothers will have a powerful motivation to keep children in a situation of dependency – since when they no longer need her, what then is her justification in life?

Indeed, as we shall explore further in Chapter 8, Firestone sees "childhood" itself as a construct, and one according to which the dependence of young people upon parents is artificially prolonged. A large part of the problem, she seems to be saying, is that both parents live vicariously through the child, instrumentalizing him or her as something through which their own lives may be validated. In fact, Firestone observes, both fathers and mothers all too often relate to the child *as property*: as an extension of their own selves, and thus as something in relation to which they have rights of ownership. The term '*family*,' she notes, 'was first used by the Romans to denote a social unit the head of which ruled over wife, children, and slaves' (67). Childhood is thus a state of subjection to the needs and the will of others. 'Childhood is hell,' writes Firestone, and 'The result is the insecure, and therefore aggressive/defensive, often obnoxious little person we call a child' (93).

Racism: the cultural pathology of the family?

For Firestone, however, the biological family with its unequal power dynamics produces not just individual but also cultural pathologies. In Firestone's analysis, as we have seen, sex oppression is the first (historical) oppression, and it has generated subsequent oppressions. According to her, the power enjoyed by men that results from women's dependence, leads them to produce further groupings of people to dominate. Sexism, therefore, causally produces both economic class and ideas of "caste" and "race." It is racism rather than economic oppression to which she dedicates sustained analysis, however, in a chapter that argues that '*Racism is sexism extended*' (97).

Firestone develops this thesis by pointing to what she calls the 'analogy' (99) between the nuclear family and American society conceived along the lines of "race." The position of black people within this family, she argues, is that of the children of the nuclear family, while the position of white men is that of the patriarchal father, and white women that of wife and mother. The psychological pathologies of domination, identification and betrayal that she argues operate within the Oedipal triangle of the family operate too, she thinks, at this larger scale. White women 'tend to trust and sympathize with black men' since they see that they too are relatively powerless; although sometimes they are led to side with the white man in his racism in the hope of thereby attaining his approval (99). The black man, however, faces the dilemma of the male child in the nuclear family: he is torn between reciprocal identification with white women and the desire to assert his own claim to patriarchal status, and often responds by 'degrading' white women in an attempt to show that, if he cannot be 'a "man" in the eyes of white society, at least he is not a woman' (101). The situation of the black woman, meanwhile, is that of being doubly exploited: by black men, who desire that she become the 'traditional passive female' as a 'negative backdrop' against which their own assertions of masculinity may be more forcefully made; and by white men, who construct her as the 'Whore' who – in contrast to the white Mother – is sufficiently degraded to be the object of sexual desire (110, 104).

The chapter in which this thesis is elaborated – entitled, 'Racism: The Sexism of the Family of Man' – is the weakest of Firestone's book. It has been rightly taken to task. Hortense Spillers, for example, has criticized Firestone for restricting her consideration of non-white women to just a single chapter, and for employing "woman" elsewhere in the book as 'a universal and unmodified noun [which] does not mean *them*.'[45] When Firestone does discuss black women, Spillers notes, she 'actually reinforces the very notions of victimization that she claims she would undo,' denying them effective political agency by 'overstating [and] misstating the black female "condition."' And Spillers takes issue with Firestone for her 'disdainfully sustained' account of the Black Nationalist Movement as 'only the last picture show of domination.' In another important critique, Angela Y. Davis has interrogated the way that Firestone's use of the Oedipal model constructs black men as 'harbor[ing] an uncontrollable desire for sexual relations with white women.'[46] Davis identifies Firestone as one of several white feminists who, 'whether innocently or consciously [have] facilitated the resurrection of the timeworn myth of the Black rapist.'

These criticisms are well-founded; and indeed, still others might be made. The chapter is startlingly presumptuous in the hugely generalizing claims it makes about what black men and women *feel*. Firestone proposes, for example, that black men feel 'an even greater bitterness' about the racism of white women as opposed to white men, since the former 'betokens a betrayal by the Mother' (99). These passages are also suggestive of a desire to exculpate white women of racism. Her claim that white women and black men 'have a special bond in oppression' which accounts for white women's support of the historical 'abolitionist movement' and the 'present black movement' presents a distortingly selective history that occludes, for example, the opposition of nineteenth-century white feminists to black men gaining the suffrage before them (98).[47] And to the extent that she does acknowledge the racism of white women she calls this an '*inauthentic* form of racism,' since it 'arises from a false class consciousness' – the illusion that the white woman's and the white man's interests are the same (99, my italics). Firestone's very thesis that racism is sexism extended in fact characterizes racism as *essentially* masculine: as originating with white men, and as constituting only a secondary phenomenon in white women.

Equally problematic, however, is that the claim that sexism is causative of racism establishes sexism as the more *urgent* oppression to be addressed. Firestone nowhere says this directly, and yet it is clearly implied, since the cause-and-effect model proposes sexism as fundamental and racism as only a symptom. The assumption of many white feminists in the 1970s that sexism took priority over racism in the emancipatory struggle may be considered an aspect of what African American feminist bell hooks has argued was a widespread, albeit sometimes unwitting, racism of the women's liberation movement itself.[48] She argues that many African American women experienced racism as being an equal if not greater force of oppression in their lives than was sexism, and that white feminists' neglect or merely token acknowledgment of this contributed to large numbers of black women becoming alienated from a feminist movement they had initially welcomed.

Unlike many other white feminists of the period, Firestone does at least attempt a substantial analysis of racism. And it is one that tries to think sexism and racism together, as interlocking oppressions: she writes that 'sex and racism are intricately interwoven' (95). In some respects, this might seem to foreshadow later theorizations of *intersectionality*, according to which systems of domination overlap and are mutually constituting, reflecting how identity itself is constructed at the juncture of different categorizations (so a US black woman experiences a racialized form of sexism, and a sexualized form of racism, for example). Theorists of intersectionality, however, do not posit one oppression as the bedrock of the others. That Firestone does, produces at least some of the profound problems with her account.

With Firestone, "sexism," or perhaps more precisely, "the sexism of the biological family," is elevated to a position where it is deemed to be the *primary* determinant of manifold social ills. It is a form of theoretical overreaching that sometimes produces reductive analyses in her text, and this is nowhere more evident than in her attempt to fit the complexities of American race relations into her rewritten Oedipal drama. It is an analysis that completely occludes, for example, the historical conditions of late-twentieth-century American racism in plantation slavery and its ongoing material and ideological effects. And the suggestion that "childhood" is the most useful lens through which to understand black American lives produces – however inadvertently – the very infantilization of black people that has been a key feature of racist ideology since slavery. As Spillers notes, 'the parental possibility does not even exist for her black characters, is not even imaginable.'

Interrogating Firestone: is the family inevitably oppressive?

Earlier in this chapter I argued that Firestone illuminates the damage wrought by *patriarchal* family structures: by families predicated on domination, and on the particular inferiority of adult women and of girls. But is Firestone correct in supposing that this is the character of *all* families that consist of the unit mother/father/child? Is it the character of nuclear families today, when the very success of second wave feminist ideas has significantly transformed that unit, with many mothers also having paid jobs, and fathers being more involved in childcare than ever before? Was it even the case in Firestone's day that *all* nuclear families were of this type – or was she describing only a particularly authoritarian and traditional version of the family?

The critique of the family is, for bell hooks, another aspect of white feminism that has alienated black American women. Black women, she argues, were likely to experience the family as being a site of resistance against racism, as much as, or more so, than as a site of gendered oppression.[49] For her, the white feminist identification of the family as the chief site of a woman's oppression was therefore yet one more example of *false universalization*, wherein white feminists falsely projected their specific experiences and interests as being the experiences and interests of all women. The British historian, Carolyn Steedman, has written of her childhood discovery that in the social world her working-class father was neither powerful nor respected – was *not* a patriarch.[50] She argues that feminist accounts of the family and family psychology that proceed from middle class experience need to reckon with what happens beyond that experiential base – when, for example, the father's authoritative status is not underwritten but instead undermined by the outside world.

The problem of a possible ahistoricism about Firestone's work resurfaces here. She accuses Freud, as we have seen, of universalizing: of thinking that neurosis must be inevitable, because he failed to see that the problem was the family, and that the family was not an immutable structure. But is Firestone herself guilty of universalizing? Is she taking just one form that the family can take – an authoritarian form, and one in which the father's status is validated by wider society – and imagining this to be the form that it always takes? Is she, as Assiter and Barrett have argued, downplaying the role of the social and historical context in shaping what the family *is*?

To repeat: Firestone thinks not. Her argument is that the biological family itself is based upon an unequal distribution of power, and therefore that in whatever form this family becomes socially institutionalized, it will inherit this fundamental inequality, and express it to some degree or another.

And this takes us to the nub of the matter. It is because Firestone thinks this that she also thinks that the only solution to women's oppression is to dismantle the family itself, by undoing the biological unit upon which it is based.

But is she right? Once again, this turns out to be a question with many parts. It is possible that Firestone is right in her critique of at least one form that the nuclear family can take – the authoritarian, patriarchal, variety; but wrong that this is the character of all nuclear families. It is also possible that she is right in her critique of the family (of some or all forms of it); but wrong in her historical account of how women's oppression originated. And it is possible that she is right in her account of how women's oppression originated, but wrong in thinking that the only solution to a biologically caused problem is the eradication of that biology.

In Chapter 8 we shall explore her proposals for dismantling the biological family, and for its replacement by a radical alternative unit for the raising of children. But first, we must consider the issues raised by her championing of technology as a means of overcoming the problems she attributes to biology.

7

The 1984 Trope

> *We are all familiar with the details of Brave New World: cold collectives, with individualism abolished, sex reduced to a mechanical act, children become robots, Big Brother intruding into every aspect of private life, rows of babies fed by impersonal machines, eugenics manipulated by the state...all emotion considered weakness, love destroyed, and so on.*
> (Firestone, p.188)

Firestone knows that part of the resistance to her proposals will result from the '"1984"' trope – the widespread cultural theme according to which any future technological society must be a dehumanized, dystopian, one (188). This fear of a technologized future is not unreasonable, she suggests, since it responds to the direction that technology has taken in the current male-dominated culture. But for her it is vitally important to distinguish between technology *per se*, and the particular uses to which technology may be put. In her analysis, the political Left's uncritical acceptance of the 1984 trope means that it has failed to make such a distinction. The result of this failure is the Left's adoption of an anti-technology, pro-nature stance that leaves the Left grievously ill-equipped to tackle either the oppression of women or the looming crisis of global over-population.

She therefore does not deny that there are particular uses of technology that are destructive: indeed, she lambasts the way that these and other technologies have been put to use to entrench forms of oppression. In terms of production, she gives the example of the conditions of modern factory work. In terms of reproduction, she cites the eagerness of the US government to dispense birth control to poor and African American women, and to women of the so-called "Third World," as a form of population control. But the purport of her discussion is to show that these are examples of the *misuse* of technologies that could be, and should be, used to further the cause of human emancipation.

Exactly what claim is being made here? One possibility is that Firestone is simply insisting on a distinction between the particular uses to which any technology is put and the potential that it may have for other kinds of usage. This seems perfectly fine. Sterilization, for example, has been used in coercive and abusive ways: as it was for many years in the US when between 100,000 and 150,000 low income women were sterilized annually under federally funded programs, with many of those forced into accepting the procedure through threats to remove their welfare benefits.[51] But this does not mean that sterilization can't be used in beneficial ways, to enable a woman or man to enjoy a sex-life without fear of unwanted pregnancy or the inconvenience of contraception. Firestone, however, often seems to want to say more than this: to say that a particular technology or perhaps even technology in general is *intrinsically* liberatory; that it is liberatory *in itself*. She explicitly avows this in the case of a range of particular technologies that were highly controversial when she was writing: 'atomic energy, fertility control, artificial reproduction, cybernation, *in themselves*, are liberating – *unless* they are *improperly* used' (179; the first and third set of italics are mine).

Is this idea about intrinsically liberating technology sustainable? Perhaps it is in a purely technical sense. Firestone's very definition of technology is that it is the 'accumulation of skills for controlling the environment [which enables] the realization of the conceivable in the possible' (155). In other words, human beings encounter a natural world that operates entirely indifferently to their needs, interests and goals, often to the detriment of human welfare. Technology, however, frees us from the necessity of these natural processes; it gives us the capacity to intervene into them not only to minimize their harmful effects, but also to harness their powers in order to achieve certain "goods" that we have conceived. In a strictly non-normative (descriptive, value-free) sense of "allowing for the breaking of natural patterns," technology can then perhaps be said to be "liberating." Firestone, however, seems precisely to mean "liberating" in the normative sense of "freeing people from oppression": technology in this account is therefore a "good" in itself. When she writes of technology being misused in order to harm, abuse, or oppress, she therefore tends to characterize such usages as being *perversions* of what technology essentially or properly is. This use, I think, is more difficult to defend (although it might have benefits as political rhetoric). Clearly, technology can be just as easily used to further as to combat oppression, so to suggest that it is intrinsically "liberating," in this normative sense, seems to be a matter of smuggling one's views about what *ought* to be the case into a purportedly objective description.

But if this is a problem with Firestone's treatment of technology, it is one that is easily addressed. Perhaps Firestone need only to have said that technology is *potentially* liberating, or can *potentially* be employed in the struggle against oppression. That, surely, is indisputable; and it is already implied in her distinction between a technology itself and the particular uses to which it is put. Firestone's affirmation of the potential of technology is, I think, an enormously valuable intervention into a form of 1960s politics that, confronted with technological horrors such as the nuclear threat, was perhaps too keen to throw the baby out with the bathwater. In the remainder of this chapter, I shall present a – qualified – defense of Firestone's position on technology by reading her work as an early example of feminist critique of science.

Two cultures

In her ambitious chapter, 'Dialectics of Cultural History,' Firestone analyses examples of *misuse* of technology as the consequence of the way that science and technology have been shaped within a male-dominated culture. Drawing upon C. P. Snow's influential Two Cultures formulation (from his 1959 Rede lecture, and then developed into a book) she argues that it is indeed the case that a schism has been created between science and the humanities, such that the two fields have become so separate as to constitute independent cultures virtually 'incomprehensible to each other' (154). For her, however, this is in turn an effect of a much deeper schism within human culture – that which is created by the sex dualism itself and the distortions in male and female psychology that arise as culture reinforces a biological inequality.

In actuality, she suggests, technological science and artistic creation ought to be compatible: they have shared origins as aspects of 'culture,' which she defines as '*the attempt by man to realize the conceivable in the possible*' (154). Uniquely among animals, human beings are able to 'project mentally states of being that do not exist at the moment' and thus are capable of experiencing a tension between the world as it is and the world as they would like it to be (154). Art and technology are both attempts at resolving this tension: art is the practice of imagining 'objects and states of being' that do not currently exist and rendering them through forms of representation (visual, linguistic, etc.); while technology is the accumulation of knowledge and skills in relation to manipulating the physical environment, in order to create such objects in reality (154). There therefore exists a dialectic between art and technology, where art anticipates in fantasy what technology is only subsequently able to produce in reality (her example is the moon landing), but where technology itself often first suggests new directions for the artistic imagination. These two modes, however (the Aesthetic and the Technological), correspond closely to female and male behaviors as they have developed in a male-dominated society; the Aesthetic Mode requires intuitiveness, introversion and dreaminess while the Technological requires a rational, objective, pragmatic response. The realm of science and technology has therefore become the almost exclusive domain of men (in contrast, women are represented in the fields of artistic production, although they are outnumbered by men and subjected to male aesthetic standards). In consequence, the familiar 'catalogue of scientific vices...duplicates, exaggerates, the catalogue of "male" vices in general,' since 'if the technological mode develops from the male principle then it follows that its practitioners would develop the warpings of the male personality in the extreme' (165). The current scientific culture of empiricism (which seeks the 'total understanding of Nature' (163)) is, as such, beset by internal contradictions. The scientist is obliged to suppress all feelings (about the impact of his discoveries, or the suffering of his experimental subjects) and to pursue scientific discovery for its own sake. Technology thus takes on 'a life of its own' – resulting in such horrors as the atomic bomb (163). The problem is that science and technology have been alienated from what for Firestone is their true goal – that of the improvement of life for (all) humanity.

Firestone's pro-science and pro-technology stance was another aspect of her thought that set her at odds with many other feminist activists at the time. It contradicted the calls prevalent within the counter culture for a "return to nature," and women had good reason to be critical of a scientific tradition that for centuries had not only excluded women from its ranks but had often also colluded in their oppression. (Victorian scientists, for example, never stopped expostulating about the female biological weaknesses that required men chivalrously to protect women from burdens they could not bear – such as voting or owning property.) On this view, Firestone's embrace of the masculine domain of science can appear as an aspect of her alleged male-identification (see Chapter 5's discussion of pregnancy).

Her views on technology have also been criticized subsequently, by those for whom she evinces a form of *technological determinism*. Mary O'Brien, for example, suggests that readers 'might well feel some misgivings about Firestone's rather naive reliance on technology,' questioning why mechanized reproduction should be expected to be liberating when mechanized production has contributed to domination and alienation.[52] Donna Haraway fears that in positing nature as the 'enemy,' Firestone 'prepare[s] for the logic of the domination of technology – the total control of now alienated bodies in a machine-determined future.'[53] These and other similar criticisms vary in their emphases, but they tend to propose that Firestone shows an unwarranted optimism about the emancipatory potential of technology; and even that she risks encouraging a kind of political quietism by suggesting that we can stand back and calmly have faith in technology's unfolding towards our liberation.

For me, however, this accusation misses its mark. As Sarah Franklin and other more recent commentators have argued, Firestone is not the naive champion of technology that she is sometimes taken to be, but a thoughtful critic of how a practice that has the potential to contribute to human freedom gets made into its opposite.[54] She is in no way suggesting that we can somehow leave it to technology to put things right. Instead, she is very clear that feminist agency is needed, and that it must take the form of women gaining direct control over reproductive technologies. In her analysis, Orwellian 1984 scenarios threaten only when "male" attitudes to science are given free rein.

She is therefore actually advancing a *feminist critique* of science; one that refuses to accept a view of science as developing neutrally or objectively and that insists instead that the course of scientific research is influenced by the values and prejudices of the wider (in this case, male-supremacist) culture of which it is a part. She argues, for example, that the technological know-how to create the female contraceptive pill existed long before the pill itself, progress in this direction having been impeded by a moral framework according to which the purpose of sex was reproduction and not pleasure – and certainly not women's pleasure. One immediate consequence of this is that it is a revolutionary imperative that women now enter the ranks of science, since only then can we expect to see an alteration in the course of technology towards the transformation of reproduction that women's freedom requires:

> It is clear by now that research in the area of reproduction is itself being impeded by cultural lag and sexual bias. The money allocated for specific kinds of research, the kinds of research done are only incidentally in the interests of women when at all...that women are excluded from science is directly responsible for the tabling of research on oral contraceptives for males. (180)

Neither celebrating science and technology uncritically, nor dismissing them, the *Dialectic* instead mounts a passionate call to the Left to give up its dangerous romanticization of nature and 'rather than breast-beating about the immorality of scientific research' focus radical energies on 'demands for control of scientific discoveries by and for the people' (179). In fact, Firestone argues, the current stage of male-dominated empiricism is only 'to culture...what the bourgeois period is to the Marxian dialectic' – a period whose manifest contradictions contain the potential to generate revolution (163). There will therefore be a cultural revolution in addition to an economic and sexual one, and it is one that will bring about 'the reintegration of the Male (Technological Mode) with the Female (Aesthetic Mode), to create an androgynous culture surpassing the highs of either' (174).

The family as anti-revolutionary force?

For Firestone, fear of technology is really fear of dehumanization: of a world in which the technocratic values of efficiency, quantification and control have supplanted the more "human" ones of kindness, nurturing and love. Because the family is so frequently held to be the site of these human values, it is perhaps unsurprising that in so many dystopian visions of a technology society, the family has been all but destroyed. Orwell's novel is a case in point. Part of the horror in *1984* is that parents have been brought to fear their own children, who at any moment might denounce them for 'thought crime.' With its former bonds of trust and loyalty now extinguished, even the family in *1984* has succumbed to the total encroachment of totalitarian state power.

Firestone in fact accepts that 'despite its oppressiveness [the family] is now the last refuge from the encroaching power of the state, a shelter that provides the little emotional warmth, privacy, and individual comfort now available' (188). She therefore needs to show that her vision of a future technological society differs from the 1984 scenario in this crucial respect, that in eliminating the family it is not seeking to destroy this last preserve of warmth and human kindness, but rather to liberate those qualities from their containment within the family unit. Her proposed alternative to the nuclear family will seek to 're-establish the female element in the outside world, to incorporate the "personal" into the "public,"' to diffuse these qualities in order to 'humanize the larger society' (191).

Again, here we see Firestone offering a more nuanced analysis than is often allowed. It is not that there is *nothing positive at all* about the nuclear family. But part of the problem is that it is asked to do *too much*. It is made to take on an impossible task – to be a sanctuary against the encroachments of an increasingly hostile world – and some of its pathologizing effects result from this. There are some interesting correspondences with writer Mark Fisher here, whose recent *Capitalist Realism* argues that 'with the public sphere under attack and the safety nets that a "Nanny State" used to provide being dismantled, the family becomes an increasingly important place of respite from the pressures of a world in which instability is a constant.'[55] Fisher points out how, ironically, one tendency within capitalism itself is towards the undermining of the family (which at the same time it relies on, to reproduce workers), as the demands of the workplace 'den[y] parents time with children, [and put] intolerable stress on couples as they become the exclusive source of affective consolation for each other' (33).

Late capitalism – both Firestone and Fisher seem to say – encourages a diminution in our sphere of affective relationships, of the number of people whom we can trust, or towards whom we have a responsibility of care, until this sphere encompasses immediate family only. The negative results are two-fold. What is consequently demanded of partners, parents or children is something that is impossible to achieve – that they will meet all of our emotional needs – and hence the family all too often becomes a site of disappointment and recrimination. And at the same time lack of emotional responsiveness to others outside this circle – lack of kindness and compassion, expressed as political support for policies against immigration, or against the redistribution of wealth – becomes reconfigured as morally permissible, even admirable.

Firestone calls this attitude the 'chauvinism that develops in the family' (177). She links it to the public lack of concern about over-population and its threat to human well-being and even survival: 'why worry about the larger social good just so long as You and Yours are "happy"'? The nuclear family gestates an 'Us-Against-Them chauvinism' (177). Blood is thicker than water. The public is a matter of abstraction. What really matters is the private and the concrete. The family functions as a form of ideological enclosure, appropriating human kindness for itself, and jealously containing it within its own borders. As Stevi

Jackson notes, Firestone's concerns here are very close to those expressed by the socialist feminist, Lynne Segal. She quotes Segal thus:

> In our intensely individualist, competitive, capitalist society, love and concern for others become inappropriate outside our very own small family groupings. Class privilege and racist exclusion are most frequently justified, by both women and men, in terms of the interests of one's own children.[56]

In contrast, Firestone asks us to consider what it would be to liberate warmth and compassion from their confinement to within the nuclear family – to humanize a dehumanizing society that is not waiting for us some time in the future but is already here. For her, what is required to achieve this *re*humanized society is the technological destruction of the nuclear family, and in the next chapter we will consider her model for how this might be achieved. But as we are doing so, we might reasonably ask whether Firestone is right to insist on this as a *necessary* condition of a rehumanized society. Is the destruction of the family really the *only* way to achieve this?

8

Revolution: To Blueprint, or Not to Blueprint?

> *The classic trap for any revolutionary is always, "What's your alternative?" But even if you could provide the interrogator with a blueprint, this does not mean he would use it: in most cases he is not sincere in wanting to know.*
> (Firestone, p.202)

The truth of these words will be evident to anyone who has followed the course of the anti-capitalist protest movements of the last decade or so. So what's your alternative? It is frequently a rhetorical question. If no answer is forthcoming, then the questioner triumphs in having revealed the pointlessness, self-indulgence, vapidity and sheer inanity of the protest. If an answer is proffered, however, then it is likely to be scoffingly dismissed as utopian, absurd, impossible. It is a manifestation of Fisher's 'capitalist realism' – the 'widespread sense that not only is capitalism the only viable political and economic system, but also that it is now impossible even to *imagine* a coherent alternative to it' (2).

Firestone was writing at a different time, before the collapse of the USSR and the global spread of capitalist economics and neoliberal ideology. In 1969 the operations of capital in the West had produced a hegemonic crisis leading to the emergence of a counter culture in which many thousands of people, especially young people, precisely did believe that an alternative was possible, and indeed perhaps even probable. Her imagined interrogator is not necessarily a representative from the older generation or the establishment, however, because what Firestone is seeking an alternative to is not only capitalism, but, as we know, biological reproduction; and she is well aware that resistance to changing *this* is as likely to come from her revolutionary confederates. The family, she says, has a uniquely powerful grasp on our imaginations, since it is the immediate structure within which our very selves are forged:

> The nature of the family unit is such that it penetrates the individual more deeply than any other social organization we have: it literally gets him 'where he lives.' I have shown how the family shapes his psyche to its structure – until ultimately, he imagines it absolute, talk of anything else striking him as perverted. (203)

As Nina Power has written, 'It's easier to imagine the end of the world than it is to imagine the death of the nuclear family.'[57]

So when Firestone comes in her final chapter to describe a possible post-revolutionary society, she does so because she recognizes the political dangers of a failure of the imagination. To be unable to imagine things being different slides easily into the belief that they *could not be* different; that, however bad things are now, they *must* be like this, and that to protest against them or call for change is just so much pissing in the wind. There is therefore a dilemma, she feels, posed by the twin dangers of not outlining an alternative – and thus inviting depoliticizing pessimism – and outlining one that will inevitably be surpassed by unfolding events. For 'any specific direction must arise organically out of the revolutionary action itself,' she says, ruling out the construction of any blueprint – if what is meant by that is a design or template to be obeyed to the letter – and insisting that radical political movements be self-critical and their objectives and strategies remain open to revision. But the force of the dilemma leads her 'to feel tempted here to make some "dangerously utopian"' concrete proposals. She stresses, however, that these are mere sketches, 'meant to stimulate thinking in fresh areas rather than to dictate the action,' and she begs her reader to 'Keep in mind that these are not meant as final answers' (203).

Unfortunately many don't. Firestone is frequently ridiculed on the basis of her revolutionary proposals, the alleged unfeasibility or undesirability of which are then treated as sufficient grounds to dismiss her work in its entirety. In the following I shall argue that some aspects of the model she draws up are not so very far removed from proposals that are in fact seriously considered, at least in radical political movements, today. One reason this is important is that Firestone's revolutionary future, seemingly dependent upon technological developments that still today seem the stuff of science fiction, can seem so far off as to invite the kind of paralyzing quietism that her sketch is intended to ward off. Artificial wombs? Robots to eliminate alienated labor? If these are a precondition of liberation, then we might seem to be so far away from them that it is hard to see what form any radical political action could take *now*, in their absence. I suggest that this problem is partially ameliorated by the existence of more concrete and currently achievable, although less dramatic courses of action, that Firestone's text points to.

The demands

While Firestone's proposals about the form a post-revolutionary society might take are tentative, the *demands* that any such society must meet are not. These are described as 'structural imperatives' (187), meaning that each is a necessary condition for the elimination of the oppressions wrought by patriarchal capitalism. She identifies four such demands:

1. *The freeing of women from the tyranny of reproduction by every means possible, and the diffusion of the child-rearing role to the society as a whole, men as well as women.* [...]
2. *The political autonomy, based on economic independence, of*

both women and children. [...]

3. *The complete integration of women and children into society.* [...]
4. *The sexual freedom of women and children.* (185–7)

For most of these demands she will identify measures that can be taken immediately and that do not rely upon the imagined development of technologies, although she typically considers such intermediate measures to be matters of reforms, which do not satisfy in full the necessity for revolutionary transformation. The imperatives can be elucidated in relation to four key headings: production, reproduction, children, and sex and sexuality.

Production

Key to emancipation will be a transformed relationship to production, and this, Firestone believes, is possible now, for the first time in history, because of *cybernation* – the replacing of human workers through advanced technologies such as computers.

Cybernation is of course in evidence all around us today, and to that extent Firestone's predictions were prophetic. When we take money out of the bank, buy goods in the supermarket or train tickets, we can do so without engaging with another human being at all. Of course, critics suggest that this contributes to a society of alienation, and workers and trade unions justifiably fear automation being used to lay people off. Firestone is not unaware of such dangers. In keeping with her "dialectical" approach to technology she sees cybernation as it develops under current economic and political structures as producing a probable worsening of conditions – albeit one that *eventually* hastens revolution. It is likely first of all, she predicts, to produce a range of 'lower rung...white collar service jobs' ('keypunch operator, computer programmer') whose transitional character will make women ('the transient labor force par excellence') the ideal employees (183). In consequence of bringing women into the labor force in unprecedented numbers, this development will, secondly, erode the male's status as 'head of the household' and 'may shake up family life and traditional sex roles... profoundly' (183). Thirdly, she forecasts that as automation replaces traditional jobs and increases unemployment there will be an increase in 'unrest of the young, the poor, the unemployed' (183).

Much of this was prescient. Firestone seems to have predicted what theorists describe as today's *feminization* of labor – the increasing numbers of women within the paid workforce since the 1970s and the increasing precarity (for both men and women) of many jobs. It is also undoubtedly the case that the displacement of the man's traditional position as sole (or main) breadwinner has brought about a restructuring of traditional sex roles, with few men in families today occupying the position of unquestioned patriarch (although Firestone would no doubt insist that this restructuring is only partial, that it does not represent true equality of men and women within the family and could not do so until reproduction itself is fundamentally transformed). Whether her prediction will be proved true, that by increasing the difficulty of obtaining employment, cybernation will make 'revolutionary ferment...become a staple' (183–4), remains to be seen. It would be hard though to contend that unemployment – and poverty and precarity *within* employment – are not hugely significant motivating factors in today's anti-capitalist movements.

The claim that cybernation is integral to emancipation is not only a claim about its ability to foment revolutionary feeling, however. The real crux of the argument is that Firestone sees cybernation as holding out the promise of *eliminating alienated labor*. On a Marxist understanding alienated labor refers to what happens to work under conditions of industrial capital, when the worker is deprived of his or her ability to determine a project for themselves and to see it through from inception to completion. Often taken to be typical of the kind of labor Marx had in mind is the situation of the factory worker – condemned to the drudgery of fitting just one part to a product as it makes its way down the production line – but the situation of today's legions of workers within the service economy fundamentally fits this description (think of the call center operator, without power to change much that a customer might be complaining about, and distrusted even to depart from their script). In Firestone's analysis the end-point of cybernation will be the taking over by machines of wearisome and unpleasant functions previously performed by people, and this will liberate people to pursue meaningful work activities of their own choosing.

As with most of her proposals, there are transitional and more distant goals. Under the 'socialism of a cybernetic economy' (210) the initial aim would be to 'redistribute drudgery equally' (211). During this transitionary period, 'while we still had a money economy' there would be for each person 'a guaranteed income from the state to take care of basic physical needs' distributed 'regardless of age, work, prestige, birth' (211). The aim, however, is to eliminate drudgery altogether: 'With the further development and wise use of machines, people could be freed from toil, "work" divorced from wages and redefined' (211). Her ultimate aim is to produce for all 'What is now found only among the elite, the pursuit of specialized interests for their own sake' (211). Under 'cybernetic communism,' a monetary economy will have been replaced by the direct allocation of resources according to need, time will have been freed up for leisure and work will have been redefined as '"play"' – meaning not that it is not serious, but that it is undertaken for reasons of its intrinsic interest to the individual who is doing it, since work that does not have this interest will have been outsourced to machines (213, 211).

Such arrangements would achieve the second of Firestone's structural demands, for the 'economic independence and self-determination of all':

> Under a cybernetic communism, even during the socialist transition, work would be divorced from wages, the ownership of the means of production in the hands of all the people, and wealth distributed on the basis of need, independent of the social value of the individual's contribution to society. (214)

How feasible is all this? Can all wearisome or unpleasant work be automated? I'm not sure that it matters if Firestone is being a little too optimistic about the capacity of cybernation to do away with work that has "social value" but no "personal value" (in other words, it needs to be done, but no one particularly likes doing it) because it seems to me that in some ways her most radical proposal occurs in relation to her supposedly transitional aim – that if unpleasant and alienating work remains, then it should be equitably shared.

Let's imagine that insufficient numbers of people enjoy being responsible for cleaning hospitals (despite whatever gadgetry exists to alleviate its burden). Then this work would be distributed between people for whom the other 90 per cent (let's say) of their working time is spent on activities that for them are personally rewarding as well as socially valuable. Somebody might object that it would be hugely wasteful of individual talent to have for example, a medical doctor, spend 10 per cent of her time doing work that "anybody could do." But the point of this arrangement would be to develop the currently under-utilized talents of those who spend 100 per cent of their working time on unpleasant labor in the direction of work that is of personal as well as social value (no doubt some current hospital cleaners, given the opportunity, would become doctors). Another benefit of this arrangement would be to change the collective value placed on such labor – to reveal to everybody through direct personal experience the care, attention, patience and effort involved in the indispensable function of keeping hospitals clean.

Alternatively, if this is too radical, what about a fundamental revaluing of different kinds of work in terms of the wages attached to them? A university certainly cannot function without cleaners, caretakers, administrators and lecturers, and possibly not without a vice chancellor. And yet the lecturer typically earns at least twice that of the caretaker, and the vice chancellor perhaps ten times as much as the lecturer. This is surely an unwarranted disparity. Perhaps, while there remains less personally rewarding work that society nonetheless needs to be done, the wages of such jobs could be increased by way of compensation; paid for, perhaps, by higher taxation of well-paid and/or personally fulfilling jobs.

While Firestone's speculations might in 1970 have looked far-fetched, the decades that have elapsed have in fact seen proposals similar to hers emerge in the arena of "serious" political and economic debate. "Technological unemployment" is the term given today to what many believe will be the increasing levels of *structural unemployment* – permanent, because produced by the fundamentals of the system itself – brought about by the automation of jobs. (Economists in fact debate whether permanent unemployment will result from automation: some argue that, as in the past, automation will produce new kinds of jobs; others counter that in this respect the future will not be like the past, since where formerly people gained work servicing the machines that had eliminated their jobs, now machines will build, service and repair other machines.)[58] This has contributed to calls for the introduction of a universal basic income (UBI) – a non-means-tested income paid by governments to citizens regardless of whether or not they are in paid employment. Such calls have come from across the political spectrum, from voices on the Left seeking to eliminate poverty and reduce inequality, as well as those on the Right, for whom a basic income promises to reduce the costs of welfare administration, or is key to enabling entrepreneurs to pursue their business ideas and to ensuring constant levels of consumer spending.[59] In the UK today, the introduction of a basic income is official Green Party policy and is being seriously considered by the Labour Party.[60] Pilot schemes have been launched in several countries and several have indicated that a basic income decoupled from one's role in production does not significantly reduce people's willingness to work (a frequent objection to UBI).[61] But even if it did, if there is to be less work available – many argue that the five-day working week is facing obsolescence – then perhaps this is a benefit rather than a problem.

As Firestone herself would be the first to observe, the potential direction of any technological development is always highly uncertain, not least because it depends on what political choices are made. Of course, the automation of work *could* lead to an even more unequal society divided between an economic elite of CEOs, corporate managers and shareholders and an unemployed underclass (the majority) suffering ever-worsening poverty. But as Firestone points out, such a prospect is by no means inevitable. What we need is a fundamental rethinking of the value of work; of the relationship between work and the resources that people receive; and to devise radically new ways of distributing work and of utilizing leisure time.

Reproduction

We know that for Firestone, however, a revolution in production will fail to bring about emancipation unless it is unaccompanied by a transformation in reproduction. What then does she think needs to happen to reproduction itself?

There are, she states, 'many degrees of this' (185). Her first call often escapes attention but is perhaps one of the most important: that reproduction cease to be considered 'the life goal of the normal individual' (206). In a culture that makes having kids a virtual imperative, this is in itself a radical demand. Still today, the family is given sacrosanct status and those who live outside this structure are often regarded with pity or distrust (the very term "childless" of course suggests *lack*; while the rejoinder "child*free*" has a testiness about it that points to the weight of ideological pressure being resisted). Firestone's first demand is to open up possibilities for non-reproductive lifestyles, and for these to be accorded equal value and respect. Two possible forms of such lifestyle, she notes, already exist: the 'single life,' in which an individual's 'social and emotional needs' are satisfied through their professional occupation; and 'living together,' in which 'two or more partners, of whatever sex, enter a non-legal sex/companionate arrangement the duration of which varies with the internal dynamics of the relationship' (205). She observes that the problem with these arrangements, however, is that they are still considered to be outside the mainstream, and are accepted only 'for brief periods in the life of the normal individual' (205). Quickly enough, demands emerge to marry and have children. There is a need, therefore, to 'broaden these options to include many more people for longer periods of their lives,' to transfer to them 'all the cultural incentives now supporting marriage' and to make such alternatives 'as common as and acceptable as marriage is today' (205). The first step in securing women's freedom, then, is to make motherhood *optional*; and it is only truly optional if the alternative is culturally valued and materially supported. But Firestone is clear that this applies as well to men, for whom fatherhood may not entail the same sacrifices as motherhood does for women, but who also need being a parent to become a choice and not a compulsion.

A second important staging-point for Firestone in freeing women from the demands of reproduction are the calls being made by feminist activists at the time for freely available contraception and abortion (see Chapter 2 and Chapter 10). But for Firestone, while these are extremely important – indeed, basic conditions of women's freedom – they comprise only some of the reproductive technologies that she sees as promising women's delivery from their biological burden. We now come, then, to the proposal for which she is most famous, and most ridiculed: that biological reproduction itself be eliminated in favor of artificial reproduction.

Firestone doubts that there is such a thing as a natural instinct for pregnancy (as opposed to an instinct for sex, the result of which may be pregnancy). Indeed, she also puts into question the existence in either men or women of an instinct for parenthood. Parenthood is often frequently genuinely desired, she admits, but she believes that this is due at least partly to the cultural superstructure that creates inducements to begin a family, and often as a 'displacement of other needs' (205). For the man, these include the opportunity to extend his ego into the world (to continue his name and lineage), and for the woman, justification of her existence in a world that otherwise attributes her little worth. It is only when such 'false motivations' have been eliminated – by valuing equally non-reproductive lifestyles and dismantling patriarchy's pathologies – that it will be possible to see the true extent of an authentic desire to raise children, which Firestone speculates would take the form of 'a simple physical desire to associate with the young' that is likely to be experienced by men just as much as women (206).

The importance of creating an alternative to traditional reproduction is, for Firestone, manifold. As we have seen in previous chapters, part of her concern is for the pregnant woman herself, and is to do with relieving women of the physical suffering, incapacitation and threat to health that accompany both pregnancy and childbirth. But her concern lies also with the child:

A mother who undergoes a nine-month pregnancy is likely to

> feel that the product of all that pain and discomfort 'belongs' to her ("To think of what I went through to have you!"). But we want to destroy this possessiveness along with its cultural reinforcements. (208)

We return here to one of the key themes of the book: that the understanding of human relationships in terms of *property* is detrimental to all those within them. To relate to a child as an aspect of one's self, to charge it with the task of representing oneself, fulfilling oneself or justifying oneself, is to establish for it an impossible burden and for oneself an inevitable disappointment. Firestone wants children to 'be loved for their own sake,' as independent and autonomous beings, not as possessions over which parents retain proprietorial control (208). And the flip side of this idea of the child as parental possession, she thinks, is that other children (those who are not *anybody's child* in particular) are relegated in their importance, their claims on the social body for protection and nurturing going insufficiently acknowledged, as responsibility for children is privatized around the biological unit of the family. Firestone suggests that to see the truth of this one need only look at the orphanage of 1960s America, which is the 'underside of the family,' in which 'unclaimed children suffer' (188–9). But it would not be difficult to find examples in our own time, in the conspiracies of silence surrounding the abuse of children in the Church or the care system, or the abuse and neglect of unaccompanied child refugees in Europe.

This sets up an unresolved equivocation in the *Dialectic*. On the one hand, Firestone claims that she is not in fact seeking *necessarily* to eliminate biological reproduction, but only to make of it one possible route among others for a woman who wishes to mother. 'At the very least, development of the option [of artificial reproduction] should make possible an honest re-examination of the ancient value of motherhood,' she says, arguing (reasonably enough) that one cannot really assess the value of biological reproduction until one has had experience of the alternatives (181). The argument about the child-as-property, however, inclines toward the conclusion that biological reproduction really cannot be maintained if women and children are to be free, since biology plays such a key role in the idea of the child as belonging to the parent. This is not only because of the aforementioned pain of labor ('to think of what I went through…') but also, Firestone seems to imply, because the notion of biological kinship is so central to this idea of the child as adjunct of the self. Firestone's goal therefore, in proposing a radically different structure within which to raise children, is to replace 'psychologically destructive genetic "parenthood"' (214) altogether.

She calls the structure that she proposes for this the *household*, rejecting the term 'extended family' as being still too tied to the old ideologies. She describes it like this:

> A group of ten or so consenting adults of varying ages could apply for a licence as a group in much the same way as a young couple today applies for a marriage licence, perhaps even undergoing some form of ritual ceremony, and then might proceed in the same way to set up house. (207)

The children raised within the household would probably be born through artificial reproduction (conceived, gestated and "birthed" in artificial wombs). It is a structure intended to facilitate collective parenting, in which responsibility for children is diffused rather than concentrated. The chief advantages of such an arrangement, for Firestone, ought by now to be clear: with biological childbearing eliminated, and responsibility for child nurture collectivized, shared across a number of adults (and older children) of both sexes, the age-old sexual division of labor and hence its 'traditional dependencies and resulting power relations' will have been eliminated (207).

Women as well as men will be able to retain a sense of their pre-parenthood identities and continue whatever projects, work or interests they previously had alongside being a parent. The child will not be considered the *possession* of any individual. Since the household exists within Firestone's cybernetic communism (with resources provided according to need) there will be no economic dependence of women upon men, and no economic precarity to threaten the stability of the unit. Indeed, Firestone holds that the household will be a more stable structure for the rearing of children than was the traditional family, since 'all participating members' will have 'entered only on the basis of personal preference' (and not through cultural or familial pressure – the 'shot gun' wedding, for example) (207). Firestone also observes that this is an arrangement that 'allows older people past their fertile years to share fully in parenthood when they so desire' (207), and she notes the likely benefits to the child of being exposed to a range of ages, perspectives and personalities. The household is to be an unprecedentedly *democratic* organization: chores (made more efficient anyway by the household's larger scale) would be distributed equally until cybernation did away with most of them; women 'would be identical under the law with men,' while children 'would no longer be "minors," under the patronage of "parents" – they would have full rights' (209).

Children

Although she cites having and raising children as the chief source of women's oppression, Firestone is not anti-children. In fact, and as Alison Jaggar and Stevi Jackson have both pointed out, she is important as one of the earliest feminist thinkers to put the interests of children firmly at the forefront of feminist politics.[62] The frequent yoking of adult women with children ("women and children first!") is not only a consequence of how women are infantilized in patriarchal culture, she believes, but also an oblique recognition of their shared condition as an oppressed class. Children, just as much as women, have everything to gain from the dismantling of the patriarchal nuclear family – and that means that in a particular but very important sense, men do as well. While Firestone never denies that men gain all kinds of benefit from their dominance in society, she also sees them as (like women) having been psychologically deformed through their upbringing in the patriarchal family; many of the male attitudes of which she is highly critical she considers as developing out of understandable adaptations on the part of the male child to his circumstances. What then might childhood be like, were those and other circumstances to be different?

Her description of the household is predicated on the assumption that under such radically transformed conditions, *childhood itself will have changed*; children will develop much more quickly the faculties now associated with adults. This accounts for what might otherwise seem to be the surprisingly short duration that she proposes for the contracted period of the household (for as long as a legal contract is necessary, which she imagines as being only a temporary stage): 'perhaps seven to ten years, or whatever was decided on as the minimal time in which children needed a stable structure in which to grow up' (207–8). This will be 'probably a much shorter period than we now imagine' (208), since childhood (in the sense of dependency) is currently *artificially extended* by adults who have an interest in doing so. Since children will be capable of rational, autonomous decisions much sooner than is presently the case, one of their rights will be 'the right of immediate transfer: if the child for any reason did not like the household into which he had been born so arbitrarily, he would be helped to transfer out' (209). One might expect the teenage years to be eventful.

But Firestone would no doubt claim that the period of frequent conflict we call *adolescence* is itself a product of the malformations of the traditional family: that without the need to detach herself from a mother or father attempting to live vicariously through her, the familiar (pun intended) crises of the teenage years would be eliminated, or at least reduced. This optimism about the future shape of childhood bears considerable weight in Firestone's proposals, as is evident in other aspects of her account of childhood in the household.

Education is an example. Compulsory schooling will have been replaced with optional '"learning centres,"' open to both children and adults wishing to develop both 'rudimentary' and 'higher' skills (211). Once again, technology plays a key facilitating role here. Firestone believes that children will *want* to learn since education will have been transformed: the need for the 'traditional book learning, the memorizing of facts' which forms the substantial part of the current school system will have been largely obviated by the 'further development of modern media for the rapid transmittal of information,' freeing up learners to concentrate on the particular skills required for whatever specialized discipline they have chosen as their focus (211). Marge Piercy, in *Woman at the Edge of Time*, re-imagines Firestone's vision of 'computer banks within easy reach' (211) as *kenners* – small computers that people carry on their arms and turn to for instant information. In 1970, when the first IBM computers cost hundreds of thousands of dollars, existed only in universities or specialized institutes where they took up most of a large room, and had a tiny fraction of the processing power of today's laptops, these ideas might well have seemed like science fiction. Now, of course, they look like anticipations of the Internet, the smartphone and the smartwatch. Quite how these information technologies will impact education is still being determined. That e-learning is all too often embraced by schools and universities as an inferior substitute to more costly face-to-face teaching, shows once again how the technologies Firestone welcomed could be used for the worse. But of course, it doesn't have to be like this. Firestone was prescient in foreseeing both that such technologies would arise, and that they would bring to 'the apparatus of culture at least as significant a change as was the printing press' (211).

Her belief that compulsion in education can be done away with is perhaps more importantly, however, a product of her belief in a child's innate curiosity. It seems unlikely, she proposes, that a child would choose not to develop her potential through the learning centers, since 'every child at first exhibits curiosity about people, things, the world in general and what makes it tick' (213). This curiosity is deadened only when 'unpleasant reality' intervenes, and 'the child learns to scale down his interests, thus becoming the average bland adult' (213).

Perhaps she is over-optimistic in suggesting that compulsion could entirely be removed from children's education. Perhaps not all learning can be made fun, although it might still be necessary, and children often cannot know what is in their long-term interests. But her comments have value in provoking reconsideration of education, and of why it should be that in its current form it seems to turn so many kids off. Governmental education policies in the UK, the US and other neoliberal societies are not focused on producing educationally rounded and critically engaged citizens, but on preparing the majority for various kinds of alienated work. As a lecturer in British higher education, all too often I encounter students whose intellectual curiosity has *survived*, rather than been *fostered by*, their experiences at school. School teachers rightly complain of having to "teach to the test," of having to dissuade kids from thinking beyond the "model" answer, and of having to subject children to a regime of constant assessment that induces almost perpetual performance anxiety (for the teachers as well as the kids). Firestone's comments try to think education in less instrumentalized terms: as a lifelong experience that makes available to everyone what is currently the preserve of the few – the chance fully to develop one's potential and to find an occupation that is personally as well as socially valuable.

A second example of her optimism occurs in relation to the child's prospective relationships with others in the household. Firestone believes that, freed from its character of property relationship, the adult-child relationship will become radically open to being determined jointly by both parties. They 'would develop just as do the best relationships today: some adults might prefer certain children over others, just as some children might prefer certain adults over others' (209). Some such relationships 'might become lifelong attachments in which the individuals concerned mutually agree to stay together' (209). But most importantly, 'all relationships would be based on love alone, uncorrupted by dependencies and resulting class inequalities' (209).

Highly controversially, the child's freedom to develop relationships of many different kinds with adults includes ones that are of a *sexual* character.

Sex and sexuality

Firestone's imagined utopia is an androgynous one. The anatomical differences that are considered to distinguish two sexes will continue to exist, but artificial reproduction and the ending of male privilege will mean that they will have become unmoored from their historical significance. 'Genital differences between human beings' will have ceased to 'matter culturally' (11). Men and women will parent or not parent on the same terms, have the same range of work and leisure pursuits open to them and the psychological differences that we today associate with the sexes will have ceased to obtain, since children of neither sex will be required to suppress any part of their personalities in accordance with gender norms. Not only will individuals be androgynous, but *culture itself* will be too, the Male (Technological) Mode and the Female (Aesthetic) Mode having achieved final dialectical synthesis in a revolution that 'create[s] an androgynous culture surpassing the highs of either cultural stream' (174).

Genital differences will have ceased to matter sexually too. The logical outcome of this psychological and cultural androgyny is that the category "sex" is unlikely to be a salient one in choice of erotic partner. Human beings will have shifted towards what Firestone calls a 'healthy transsexuality' (54) or '*pansexuality*' (11) that is precisely not a matter of homosexuality or even bisexuality.

In fact, the *Dialectic* is littered with some rather discomfiting comments about homosexuality. Homosexual men are 'often misogynists of the worst order' she proclaims casually, in a discussion of how women are subjected to male-defined standards of beauty (136). But they are also the 'extreme casualties of the system of obstructed sexuality that develops within the family' (53). She does describe homosexuality as being 'at present...as limited and sick as our heterosexuality,' meaning that *any* sexuality that is predicated upon exclusion of erotic attachments based upon the category of anatomical sex is the product of sexual repression as it operates in the patriarchal family (53). But while she is aiming at a sexual liberation that will render obsolete the very categories "heterosexuality" or "homosexuality," her frequent depiction of male homosexuals as either offenders or victims (at a time when gay sexual activity was illegal in the majority of US states, and homosexuality still an official mental disorder for the American Psychiatric Association) trades thoughtlessly in homophobic terms of reference, and acknowledges nothing of the progressive work being done by gay political activists at the time.

Firestone's *pansexuality* is about even more than the undoing of sexual identities, however. Here, Freud comes back into the picture, with Firestone drawing upon his idea of *polymorphous perversity*. In Freudian theory, this refers to the condition characteristic of infantile sexuality (present more or less from birth) in which the infant's entire body is capable of functioning as what Freud calls an 'erotogenic zone' – of producing pleasurable sensations when stimulated – and in which members of either sex may form the objects of his most intense yearning. This multi-formed (hence 'polymorphous') sexuality is "perverse" only from the perspective of what Victorian society considers "normal" sexuality – genital, heterosexual sex – and Freud intends no denigration in using this term. But the infant's initial disorderly sexuality will undergo a process of territorialization, of classification into sexual and non-sexual zones, as the child's body develops and as he or she enacts the repressions that society requires. Thus it is that most people subscribe to what is for Freud the entirely mistaken view of sex as having predominantly to do with genitals, with orgasm and as beginning only at puberty.

Firestone wants to undo this process, to obviate the need for the sexual repressions that produce fixed sexual identities and to return people to a 'more natural polymorphous sexuality' (215). This will change our very conception of the "sexual," since, for Firestone, the very binary distinction between sexual and non-sexual feeling is a product of repression; what happens when as children we are taught that certain of our passionate attachments must preclude physical intimacy and pleasure. With emotional and physical affection fully reintegrated, 'non-sexual friendship (Freud's 'aim-inhibited' love)' would disappear, since 'all close relationships would include the physical' (215). Sexual activity would form only one component within a 'total' response to another person, and as such, Firestone believes, exclusive desire for one sex or another would be unlikely, as this could only proceed from making a 'purely physical factor...decisive' (although she does speculate, rather oddly, that 'all else being equal' people might prefer the opposite sex for reasons of 'sheer physical fit') (54, 215). With this diffusion of physical/emotional love across all close relationships, monogamy would of course also disappear. She writes that experiments with "free love" have failed only because the institutions that bring about sexual possessiveness – primarily the family – have not been abolished.

What is it exactly that is supposed to make possible this revolution in sexual and emotional relationships? Firestone's answer is that the replacement of the family by the household will remove the need for the incest taboo. In Freud's theory, this is the taboo that instigates the 'latency period': the huge wave of repression that takes place in the (male) child at about the age of six, and that obliterates conscious sexual desire so completely that the child appears then to be without sexual feeling until puberty. The child comes up against the incest taboo when he realizes that his passionate love for his mother is so disapproved of by his father that he is threatened with a punishment that is presented, or that he interprets, as a threat to his genitals (castration). In "normal" development the child resolves this dilemma by identifying with his father, which means internalizing the father's prohibition as his own, and through this he is able to repress the sexual part of his feeling for his mother and transform this into a current of "pure" affection. In Firestone's rereading of Freud, as we have seen, the boy's renunciation of his mother has less to do with phantasmatic fears for his penis and more to do with paternal inducements to side with him in return for access to the world of 'travel and adventure' (53). It is nonetheless the case, she thinks, that this and subsequent repressions required to bring about 'sexual normality' come at the cost of great 'psychological penalties' and make 'a totally fulfilled sexuality impossible for anyone' (52–3). The reason for the widespread belief that women have a lower sex-drive than men is, she thinks, that the sexual role to which women are pressurized to conform is particularly unsatisfying; although she notes that damaged male sexuality is more harmful – 'the confusion of sexuality with power, hurts others' (53).

Firestone believes that the incest taboo will have 'lost its function' within the household, since this being such a 'transient social form' there would not be 'the dangers of inbreeding' (215).

However, the gist of her discussion of this taboo is not to see it as a biological imperative, but as a symptom 'of the power psychology created by the family' (51) as the patriarchal father claims sexual possession of the mother for himself, obliging his sons to go outside the family to satisfy their own erotic needs. In a reproductive unit that is not premised upon relationships of property in others, this psychological imperative will not apply.

Thus it is that Firestone reaches her highly controversial views on children and sexuality. Without the incest taboo, there will be no repression of childhood sexuality and no dichotomizing of erotic and affectionate feelings – the child will be free to develop relationships that include as much of a physical and erotic dimension as it desires. These may be exclusively with other children, she speculates, but 'if not, if he should choose to relate sexuality [sic] to adults, even if he should happen to pick his own genetic mother, there would be no *a priori* reason for her to reject his sexual advances' (215). Firestone further writes that

> Relations with children would include as much genital sex as the child was capable of – probably considerably more than we now believe – but because genital sex would no longer be the central focus of the relationship, lack of orgasm would not present a serious problem. (215)

The sexual abuse of children is total anathema to what Firestone is proposing. Nor is she advocating what today might be called the *sexualization* of children, since what she aims to enable is precisely not the imposition upon children of a sexual awareness that comes from adults, but the expression of what she believes is the sexual feeling that is within children themselves, and only currently repressed. But are not these comments nonetheless dangerously naive, opening the door to the sexual exploitation of children by adults? Firestone would no doubt want to say that this could not happen; that even if a pedophilic drive should still exist after the family and its sexual deformations had been eliminated (which she would presumably doubt), the household's diffused structure of parenting would mean that there were always other adults to whom a child being mistreated could turn for protection. Yet this assumes the capacity of the child to recognize and identify an adult's behavior towards them as abusive rather than loving.

Here a problem I have already noted reappears in its starkest form, and it has to do with a malleability required of childhood by some of Firestone's proposals. It is one thing to say – as Firestone does in her chapter, 'Down with Childhood' – that childhood is historically variable; that how children are thought about, and even what they are actually like, varies under different social conditions. This is, I think, unquestionably true. But it is another to treat childhood as being infinitely plastic, as though there were nothing intrinsic to the early years of human life that constrains what it can become. Firestone does not in fact say only that childhood will be transformed. She says that the very concept will be 'abolished' (214). She means that "childhood" is simply the sentimentalizing name given to a state of dependence that is artificially imposed upon the very young and that can, under the right circumstances, be thrown off. Her liberated children are in fact in many ways indistinguishable from adults, possessing a capacity for insight into their own long-term interests and judgment as to the motivations and characters of others that many adults struggle to achieve. If one does not trust that even under the radically transformed circumstances of the household children could become like this, then some of her proposals look not only unfeasible, but in some cases, complacently, dangerously, so.

Perhaps a similar problem relates to her discussion of sexuality more generally. It's important to note, first of all, what liberated sexuality for her is *not*. It is not what we have today, despite what might seem to be an acceptance of a vastly wider range of sexual attitudes and practices. It is not Ann Summers on the high street and sex toys at hen parties. It is not Internet pornography, and the desire of young males for a girlfriend who is up for everything he has seen on the web. It is not pole dancing clubs, and Girls Gone Wild, and naked dating shows on TV. These things, which feminist writers today including Ariel Levy and Natasha Walter describe as the *pornification* of culture,[63] would have been considered by Firestone to fall under the category of *repressive de-sublimation*. This concept, taken from Herbert Marcuse, Firestone describes as what happens when a limited revolution is won within the context of a still-existent repressive structure: 'only a more sophisticated repression can result' (57). Sexuality is released into '"formerly tabooed dimensions and relations"' but in a manner that is only the appearance of a loosening of social control (57). The sexual feeling that seems to have been made free is in reality being put to work – in the interests of the economy, in the interests of male domination.

Firestone is seeking, by contrast, a genuine undoing of sexuality from its commodification. In common with many thinkers of the second wave, she seems to hold that there is an authentic sexuality waiting to be rediscovered. A term she frequently uses to characterize the sexualities of her day is *deformation*, suggesting a true form that is bent out of shape by the distorting forces of patriarchal capitalism. Later in the 1970s, French theorist Michel Foucault would offer a conceptualization of sexuality and power that if true would make such an idea – while perhaps attractive – difficult to maintain. For Foucault, sexuality is not repressed by power; it does not await liberation from it. Rather, sexuality is always the product of power relations: pleasure and desire are made possible by power's operations. There is no sexuality that pre-exists power, or is conceivable outside of it.[64]

Firestone, however, grounds her ideas of a sexuality that is to be liberated in nature: 'in our new society, humanity could finally revert to its natural polymorphous sexuality – all forms of sexuality would be allowed and indulged' (187). Ironically, and as Nina Power also notes, despite the *Dialectic*'s powerful arguments against the romanticization of nature, wherever 'natural' appears in the context of discussing sexuality, it has taken on positive connotations.[65] Since Firestone's utopia is one in which domination has been eliminated from human relationships, it must be concluded that the 'all forms of sexuality' that fall into this natural sexuality do not include ones that involve a desire to possess, control or inflict suffering; that such desires are not native to human sexuality but are the products of the distorting influence of the patriarchal family. Not only is the sexuality to which Firestone wishes to return *natural*, it is also fundamentally *nice*.

The problem is that this is difficult for Firestone to sustain. For Freud, polymorphous perversity is the condition of very young infants only; it is not a condition to which adults could or would want to return. The polymorphously perverse infant is such a chaos of drives and impulses that it cannot really be called a "person" at all. The infant is totally self-centered and among its impulses are ones aimed at hurting and destroying. It is only when what Freud calls the 'reality principle' sets in that the infant becomes capable of reckoning with an external reality and deferring gratification. A degree of repression is for Freud, necessary for the development of consciousness and rationality; it is certainly necessary for it to become a social being. Recognizing this, Marcuse had distinguished between what he called *basic* and *surplus repression*. Basic repression refers to the taming and controlling of primal drives that is needed for the emergence of a person at all, while surplus repression refers to any additional repressions that are only contingently required by a particular social formation (for example, the repression of homosexuality in some cultures). But Firestone makes no use of such a distinction. As Power again notes, 'she refuses the argument that repression plays a necessary role in the creation of culture.'[66] Or even, she simply ignores it. For her, repression can just be done away with, leaving a human being revealed as good in its nakedness. What this confidence about the innocence of polymorphous sexuality does is to discredit the possibility that there may exist something less knowable, less predictable or less likeable in natural human sexuality (if such a thing even exists). It is as if here in the *Dialectic* Firestone is not able quite to escape the clutches of hippie-Rousseaueanism: there is something good about the human being in his natural state, if only we could dismantle the bad institutions that corrupt him.

Finally, what is to be made of Firestone's vision of an androgynous society? On the one hand, freeing people from expectations based upon the details of their anatomy is surely to be welcomed. Feminist theory has long sought to distinguish between *gender* – psychological characteristics and behaviors – and biological *sex*, arguing that sex does not determine the gender identity of an individual (that this is instead acquired through processes of cultural conditioning). Simone de Beauvoir, among many others, is offering a version of this sex/gender distinction. In the 1990s, queer theorist Judith Butler would come to challenge even the sex/gender distinction for ceding too much to an idea of "sex" as belonging to a realm of pure biology. In fact, Butler pointed out, "biological sex" presents not a binary but a field of physical variations in terms of the presence of external and internal sexual organs, chromosomal makeup and hormonal levels. For Butler, that we must interpret this empirical variation into two distinct sexes shows that concepts such as "sex," "biology" or "nature" are themselves always being shaped by cultural values and frameworks. Today, many, especially younger people, are rejecting notions of binary sex or gender. In a recent survey half of US millennials agreed that gender isn't limited to male and female.[67] This "gender fluid generation" is increasingly comfortable with identifications such as transfeminine or transmasculine, agender, androgynous, non-binary, queer trans or multigender. Facebook and other social media allow for self-identification along these lines, while acceptance of gender-neutral bathrooms and pronouns ("they") is growing.

This picture differs somewhat, however, from the ideal of androgyny that Firestone advocates. Where gender fluidity suggests a proliferation of differences and the potential for an individual to shift between these, androgyny can, potentially at least, involve the annihilation of differences – homogeneity, not heterogeneity. Though it is hard to know for sure that this is what Firestone proposes, since her references to androgyny are fairly limited, it is certainly suggested by her dialectical schema, which looks forward to a teleological endpoint in which the sex difference that has been the driver of history is eliminated in a final merger of masculine and feminine principles.

Why does this matter? For one thing, the androgynous vision can look like what Rosalind Delmar has called a 'counsel of despair.'[68] It is as if for Firestone sexual difference has proven so destructive, so traumatic, that there is no hope for a healthy, happy, variant of it – the only thing to do is to eliminate sexual difference entirely. The suggestion of "somatophobia" – understood by Spelman as the thought that it would be better 'if we were not embodied' – perhaps arises once again here.[69] For another, such an elimination of difference quickly becomes itself repressive, albeit unwittingly. What would happen, for example, in a society where androgyny were the norm, to individuals who identified strongly with some particular version of traditional gender, who wished to identify and be recognized for their femaleness or maleness? It is a question that arises with particular acuteness in light of Firestone's acknowledgment that there might continue to be anatomical women who wish to reproduce the old-fashioned way, with pregnancy bump, and morning sickness, and birth, and blood and pain. How might such women be regarded and treated, given the threat to social harmony that Firestone's theory suggests this could entail?

Conclusion

The Dialectic of Sex ends here, with the conclusion of the outline of Firestone's 'very rough plan' for the social arrangements that might follow after feminist revolution (216). Despite its roughness, it is a design that she believes fulfills each of her revolutionary demands. Women will have been freed from the tyranny of reproduction by the development of technology and the diffusing of child-rearing responsibilities across society. Economic independence and self-determination will have been gained for all, with economic classes abolished and all people given the opportunity to flourish. Women and children will have been integrated into society as a whole, with children's dependence upon adults, and women's (mothers') dependence upon men, rendered obsolete. Sexual feeling will have been freed from the constricted and often destructive channels into which it has been forced by the patriarchal nuclear family.

Her proposals are certainly open to criticism. There is a pronounced tendency in her writing – understandable, in light of the urgent task of enthusing her readers about a possible alternative – to downplay the difficulties and the complications of the new arrangements that she describes; to wish to put to bed too quickly the potential, even in this more compassionate, more just, more humane society, for ongoing social antagonisms.

But while I have pointed to just some of the possible problems with her proposals, it is not that I think Firestone wrong to imagine a society that is radically different, and better. On the contrary, we need such imaginings; today, if anything, more so even than in Firestone's day. While she wants us to take her historical, causal, account of women's oppression *literally* – as what really happened – her utopian future is, I think, offered in the spirit of a 'literary image' of future possibilities (203). It is an attempt to do what good science fiction does: to escape the imaginative fetters of the present and conceptualize how things could be otherwise. It is, as Gillian Howie contends, a 'hypothetical political construction,' and a potentially 'world-changing fiction.'[70]

9

IVF, Egg-Freezing, Surrogacy

A pregnant woman will never simply be a container.
(Mia Fahlén and Gertrud Åström[71])

In 1970, Firestone was writing on the cusp of technological and political developments that promised to transform reproduction. Indeed, just eight years later, in 1978, the first "test tube baby," Louise Brown, was born in Oldham, England. While her birth attracted controversy about "Frankenstein babies" and her parents – Brown has recently revealed – received hate mail, there have today been an estimated 6 million IVF babies born around the world,[72] and the procedure's widespread acceptance as a routine treatment reveals just how conventional and contingent are judgments as to any particular technology being "unnatural."

So where are we today, nearly 50 years after the original publication of *The Dialectic of Sex*? Have subsequent events borne out Firestone's claims about the revolutionary potential of new reproductive technologies? And if they have not done so, or have done so only partially or ambivalently, then to what ought this to be attributed? Would it show that Firestone had, after all, always been hopelessly over-optimistic about the potential of such technologies? Or might there be a different explanation?

For me, the great value of Firestone's work is precisely that it calls for such an interrogation. In identifying the historically causal origin of women's oppression in their reproductive functions, it compels us to ask, "To what extent do women's roles in procreation continue to contribute to their unequal status today?" But this is still the case, I think, even if one does not accept Firestone's causal story: even if one were to think, for example, that she wrongly identifies as biological caused inequalities that are actually social or cultural in origin. For it would be hard to deny that reproduction *is* a major site of gender injustice today, however one understands that to have come about. Laurie Penny, for example, has written powerfully about what she calls 'reproductive tyranny,' about the 'obscenity of living in a culture that tries to stamp itself all over one's womb and clamp itself around one's ovaries and shame XX-genotype women for owning bodies that can create new life.'[73] Recognizing this, we are immediately confronted by the questions that Firestone was asking. These are questions about *how* women's role in the bearing and raising of children impacts upon their participation in other areas of life; with *what* consequences; and about *who* is in control of a woman's reproductive capabilities.

My aim in this chapter and the next is to sketch out the coordinates of a *Firestonian* analysis of some of the major reproductive issues facing us today. There are many important issues that I do not have space to develop: for example, the sharing of childcare within heterosexual relationships, or the impact of women's continued greater domestic and childcare responsibilities on the gender pay gap and on the under-representation of women in many areas of public life. Instead, the present chapter will focus upon egg-freezing, IVF, and surrogacy, as instances of the kind of technological intervention into reproduction that Firestone was interested in. The following chapter will examine the situation in the US with respect to contraception and abortion rights, in the face of the looming horror of a Trump presidency. A key premise of my analysis in both chapters is that any investigation of these technologies must locate them squarely within their context: that of global capitalism, the spread of neoliberal ideology, and, I argue, the continuing grasp upon our imaginations of ideologies of motherhood and the nuclear family.

That we are still in thrall to these ideologies is clear. Firestone protested, in 1970, against the marginalization of non-reproductive lifestyles, the idea of motherhood as central to women's fulfillment, and a sentimentalization of maternity that conditioned what was permissible to feel, think or say about this experience. In 2016, the hegemony of parenthood continued undaunted, and although non-reproductive men are also the targets of this, the notion of a childless woman as a failure is particularly pernicious. British politician Andrea Leadsom provided an example when she presented herself as a better candidate for Prime Minister than the childless Theresa May, on the grounds that "being a mum means you have a real stake in the future of our country, a tangible stake."[74]

But the other side of this prejudice against women who are not mothers is the continuing intrusion into the privacy and autonomy of those who are. On becoming pregnant, a woman is subjected to a clear set of messages that communicate that her body is no longer her own and that she is the vessel for something more important than herself. US feminist Jessica Valenti has written on this, pointing out how an over-valuing of motherhood in women leads to a devaluing of their human status.[75] Among her many compelling examples are the erasure of women themselves from how pregnancy is often pictured (literally, as her interviewee Professor Rebecca Kukla points out, in the case of the pregnancy book *What to Expect When You're Expecting*, which begins each chapter with an image of a pregnancy bump minus the woman's head, arms and legs), and in what she identifies as the decentering of maternal health that occurs when fetal health is prioritized – a tendency of healthcare providers that finds its complement in the media celebration of mothers who quite literally give up their lives for their children (by refusing cancer treatment while pregnant, for example). Valenti is surely right. This diminishing of pregnant women's status is evident in a range of social practices: from the apparent benignity of uninvited touching of a pregnant woman's stomach; to the issuing of advice from mere acquaintances or even strangers on what she eats, drinks or smokes, or whether or how she exercises; to the vocal condemnation she may face should she decline to conform to expectations. And it is nowhere more clear than in today's US, in the state surveillance that is targeted especially at poorer income women and women of color, as Chapter 10 will explore.

Egg-freezing

Developed initially for people with severe medical conditions – such as a cancer patient whose treatment would threaten their future fertility – egg-freezing has in the 2010s been heavily marketed at healthy women in their twenties and thirties as "fertility insurance." Egg-freezing allows women to store healthy eggs with the prospect of later fertilizing these and beginning a pregnancy at a time in their lives when they feel ready to have children, but when their natural fertility will have declined. It is therefore championed as a way of putting women in control of their reproductive lives, of freeing them from the biological constraints of women's reproductive lifetime, and allowing them the option otherwise restricted to men of becoming a genetic parent later in life. For Professor Geeta Nargund, Medical Director of CREATE Fertility, Europe's largest IVF clinic, egg-freezing promises to free women from what she calls – in what sounds a very Firestonian formulation – 'nature's gender inequality.'[76] It has been lauded by some as promising to revolutionize women's lives on the scale of the contraceptive pill. But how much freedom does this really offer women? And is there another side to its potential benefits?

The first thing to note is that this technology does not free women from the "burden" of procreation, as Firestone had wanted, but in many ways intensifies it. It shares this with IVF – and egg-freezing is, in essence, the first stage of that procedure.[77] Over a number of weeks a woman must self-inject a cocktail of hormones into her abdomen to stimulate massive follicle growth and egg production, before these eggs are "harvested" through draining the ovaries while she is under sedation or anesthetic. The high levels of the hormones involved can cause unpleasant physical symptoms such as nausea and fatigue, and contribute to emotional disturbance and mood swings. The egg-collection process can produce abdominal pains lasting for days, and in the week following collection there is a risk of Ovarian Hyperstimulation Syndrome (OHSS), in which the ovaries produce too many eggs following egg-collection, leading to symptoms including pain, bloating, nausea and vomiting, and for which a woman may need to be hospitalized.

Nor is there any guarantee of success. From any batch of harvested eggs (doctors usually aim to collect 14 or more), there is no guarantee that any will make it through the process of freeze-thawing, fertilization and implantation. The UK Human Fertilisation and Embryology Association (HFEA) states that up to December 2012, approximately 580 embryos had been created from frozen eggs, and transferred to women in 160 cycles, from which resulted 20 live births.[78] Although a new procedure for flash-freezing eggs ("vitrification") should decrease damage to eggs caused by the freezing process, it remains the case that from any one cycle of egg-collection there is a significant chance of no viable embryo resulting; anecdotally, many women report undergoing several cycles without success. Nor is this process cheap. EggBanxx, one of the largest "providers," states that the process of follicle stimulation, retrieval and vitrification can cost between $8500 and $18,000.[79] In the UK, prices for one cycle average at around £4000, with additional costs of approximately £300 per year for egg storage. Critics of the marketing of egg-freezing to healthy women say that the fertility industry is downplaying the potential risks – physical, emotional and financial – of the procedure, and inducing false confidence about the capacity to control one's fertility.

As with IVF, we need to distinguish the technology itself from the industry, and from the wider economic and cultural context in which the industry exists. The *technology* of egg-freezing certainly has the value of increasing reproductive options in the face of "natural" constraints. That there is a "natural" limit on women's child-bearing years that is not so in the case of men is the cause of much heartbreak for older women or couples wishing to conceive. As Firestone insists, the "natural" is not a human value: there is no reason why *this* limit imposed by human biology should be accepted, any more than there is that a failed organ such as a heart should mean the death of the person concerned. Egg-freezing also holds out the possibility for trans men of becoming a biological parent after sex-realignment surgery (although such a possibility currently requires use of a surrogate; more on this below). From this perspective, egg-freezing indeed looks like an instance of human control over nature, of the use of technology to liberate people from natural constraints, that Firestone would welcome.

Yet the extent to which *liberation* is really at stake becomes questionable once one interrogates the actual circumstances that produce egg-freezing's popularity. Why are healthy younger women choosing to avoid starting a family until later in life, when their "natural" fertility may have drastically declined? This is a difficult question to raise, since it is also the question asked by moral conservatives – albeit often rhetorically, their answer being that it is feminism that is to blame for making women think they can "have it all." But the Left must ask this question too, from the perspective of the structural forces driving this phenomenon. Student debt, low wages, rising housing costs, the insecurity of renting, the casualization of employment, the erosion of welfare and social security: these all contribute to an experience of precarity that is leading young people to delay having children, even while they fear for their future fertility.[80] While Apple and Facebook have been praised by some for including egg-freezing among their perk options for female employees, this prompts the question as to whether the real problem is workplace practices that dissuade people from having children until they feel relatively secure in their positions. And while some women report that it is not lack of housing, or secure employment, but lack of a relationship that leads them to consider egg-freezing, once again, the question is prompted: what are the factors (economic, social and cultural) that make the prospect of parenting outside of the traditional, monogamous couple, so very daunting? And why are single parent or parental couple the only imaginable possibilities?

Firestone's call was not only for the technological transformation of reproduction, but also for its social transfiguration. In her proposed society, individuals who wished to parent would be supported collectively, through the provision of the resources that both they and the child required. This is anathema to today's neoliberal world, where the constant refrain is that only the irredeemably feckless have children for whose provision they may need to call on social support. Constantly subjected to the "parenthood-as-destiny" mantra, young people are simultaneously told that they must wait until they can afford children, even as employment and social security policies push this "when" endlessly back into a horizon that is never reached. Into this mix of anxiety and confusion come the private fertility clinics, promising the small minority of women who can afford it the prospect of solving this problem through an unreliable, invasive and potentially traumatic medical procedure. Since egg-freezing is, as journalist Viv Groskop has noted, essentially delayed IVF that may never be needed,[81] its commercialization represents an extension and normalization of IVF as a procedure for *all* women, not only for those who are actually encountering difficulties with fertility. In so doing, it helps to create and promote a culture in which women, even in the earliest years of adulthood, are asked to relate to themselves as baby-incubators, to carry with them a heightened sense of the importance of biological motherhood and of their own dwindling capacity for it (thereby dovetailing with the US pre-pregnancy movement[82]). Far from freeing women from their biological designation as the reproducers of the species, this looks very much like a reinforcement of it.

The fertility industry, through its control of IVF, already benefits from the hegemony of parenthood, and the valorization of *genetic* parenthood in particular; its "miracle baby" narrative works to obscure what is the reality for all too many of its customers: repeated miscarriage, spiraling debt, relationship breakdown and depressive illness. IVF clinics have been subjected to criticism for offering patients costly add-on treatments for which there is little or no evidence base.[83] Given the clear potential for exploitation of often vulnerable, even desperate, people, critics have called for increasing regulation of an industry that is allegedly all too often happy to overcharge clients while providing them with inadequate information about treatments, their risks and likely outcomes.[84]

But what if the problem is more fundamental still? What if it is not insufficient regulation, but the very fact that these reproductive technologies are controlled by corporate interests at all? Firestone called for intervention into biological reproduction in the interests of people themselves, not of profit creation. We need urgently to make the "utopian" move of thinking reproductive technologies outside of this framework, and in concert with an analysis of the social, cultural and economic forces that inhibit those who wish to have and to raise children from doing so. Until such time, the fertility industry has found in egg-freezing a highly lucrative new product.

Surrogacy

Commercial surrogacy hit the headlines in 2014 with the case of Baby Gammy. Born to Pattaramon Chanbua, a young Thai woman who had entered into a surrogacy agreement with an

Australian couple, Gammy was left with his mother when the couple returned to Australia with his twin sister, Pipah. Chanbua alleges that his surrogate parents rejected Gammy when they discovered he had Down's Syndrome and a heart condition, although the couple deny this, claiming that they had sincerely believed that the surrogate had chosen to keep the child. Despite the economic burden of an unsought-for child, who has needs that she is financially ill-equipped to meet, Chanbua has undertaken to raise Gammy as her own son; and following revelations that one of the Australian couple had been jailed previously for child sexual offenses, has campaigned for custody of his twin sister. This application was turned down by an Australian judge in 2016, who found that Pipah faced "only a very low risk of...being abused," and ruled that the story of the couple's abandonment of Gammy was "untrue." The misunderstanding was not surprising, however, "when a woman's body is rented for the benefit of others."[85]

Non-sexual surrogacy, of course, is made possible by the technological advances that Firestone had looked forward to in 1970. There are different forms, but it typically involves artificial insemination (in cases where the surrogate is also the genetic mother), or egg-harvesting (from the genetic mother or a donor), in vitro fertilization and implantation. Like conventional IVF and egg-freezing, it promises to liberate people from the constraints of biology. It is often a course taken by heterosexual couples for whom IVF and other procedures have failed, or where a woman has a medical condition that means that pregnancy cannot be safely undertaken. For gay men, a surrogate may be the only way of having a child genetically related to one of the partners. But like IVF and egg-freezing, surrogacy does not alleviate women of what is for Firestone the burden of physical reproduction. Rather, it *transfers* it. And in the case of commercial surrogacy, it generally transfers it to a woman who is economically disadvantaged, with all the risks to physical health that poverty entails.

Commercial surrogacy requires economic inequality to be viable. And in most instances it requires the *vast* economic inequalities that are produced by globalized capitalism. Pattaramon Chanbua, for example, said of her decision to become a surrogate that the A$16,000 "that was offered was a lot for me. In my mind, with that money, one, we can educate my children; two, we can repay our debt."[86] In other countries of the non-Western world where commercial overseas surrogacy is legal, surrogacy may offer a solution (however transiently) to unemployment, destitution or hunger. Although the payment for surrogacy might equal several times the yearly wage that the surrogate could hope to earn, it is likely to be a fraction of the cost of surrogacy in one of the Western countries where it is legal.

Fertility tourism represents the transformation of women and children into commodities on the free market. This is what was disclosed in particularly glaring fashion by what had *apparently* happened in the case of baby Gammy. Gammy's prospective parents had, it seemed, returned goods they had purchased that they considered faulty. That human beings – Gammy himself, and the woman left to care for him – could be so instrumentalized, so reduced to objects in the designs of others – provoked reactions of horror around the world. But would not this reaction be appropriate even if it were to be proven that, as they assert, the Australian couple did *not* do this, did not leave Gammy because they did not want him, because he was disabled?

That such surrogacy involves the commoditization of women is evident from their treatment at the hands of the surrogacy industry. As journalist Divya Gupta reports, for instance, the Akanksha clinic in Gujarat houses its surrogates in a house located close to the clinic: 'The women enjoy the rest and care they may not have had during their own pregnancies,' notes Gupta, 'but are confined to the house for the whole pregnancy.

Their families can visit on Sundays but the surrogates only leave the premises for medical checkups or if there is a family emergency.'[87] In her own investigation of Indian surrogacy, journalist Julie Bindel describes how a clinic tells her it is better for the women's mental health to remain in their own homes, but that they will move a surrogate into the clinic's hostel if Bindel pays enough. She reports on how the women are told 'what and when to eat and drink'; on the prevalence of women being coerced into surrogacy by abusive husbands and pimps; on how it is common practice to use two women as surrogates for one commissioning couple, and if both become pregnant, to give one of these women an abortion. She describes the system of 'sex, race and class supremacy' that the industry is built upon, which sees poorer women hired as surrogates, but younger, better educated women as egg donors: and where commissioning parents can choose surrogates from a catalogue, and select donors at £2500 to £3000 for a Caucasian, £1000 for an Indian. Though the clinics will not reveal how much a surrogate is paid, it is clearly but a fraction of what the clinic obtains through its use of her body.

Such are the horrors of overseas surrogacy that some claim this is an argument for commercial surrogacy's legalization in the West, where, it is argued, the practice can be regulated and ethical standards maintained. Commercial surrogacy, as opposed to "altruistic" surrogacy, where the surrogate does not benefit financially, is illegal in much of Europe and in Australia (despite calls in 2014 for the ban to be lifted). Its legality varies across the US, with bans in the majority of states but with some exceptions, such as California, where contracts for gestational surrogacy (where the surrogate is not genetically related) are regularly enforced and where the intended parents are legally recognized, before birth, as the parents of the child. In the UK, while commercial surrogacy is formally banned, it is legal for a surrogate to receive "reasonable expenses," which may include compensation for lost earnings. Proponents of legalizing surrogacy argue that surrogacy provides a vital service, and that while the women who become surrogates are usually not motivated primarily by money, but by the altruistic desire to make parenthood possible for childless people, surrogates should be able to receive financial compensation for sacrifices entailed.

But can any form of commercial surrogacy exist without encouraging the commodification of women and children, no matter what the intentions of those involved? Firestone is particularly helpful here. We have seen in earlier chapters that key to her analysis is an argument against *property in other people*; against the idea, that is, of having rights of ownership over other persons. It is on this basis that she objects to marriage, which, she says, has its origins in male ownership of women, in the contracting of the sexual and domestic labor of women in exchange for their financial support. The Victorian marriage contract, she shows, was only apparently freely undertaken since Victorian women, lacking the means to provide for themselves, were without viable alternatives. In a situation of radical inequality, Firestone shows, such a contract is thus coerced, not voluntary. It is an argument that voids all liberal defenses of overseas commercial surrogacy as being legitimated through the woman's *choice*. But even in the context of the wealthy countries of the West, there remain stark inequalities of resources and opportunities. Is it really possible to pay for rights of use of the interior spaces of a woman's body without feeling that that body has somehow become *yours to use*? Even if some individuals can manage this, can the use of women's bodies in this way be normalized, and legalized, without contributing to a moral climate in which women are more likely to be perceived in this way?[88] Can there be a society in which it is accepted that privileged women pay for use of the bodies of less privileged ones, and in which this is justly called "reproductive freedom"?

But Firestone also points to the potentially damaging

consequences of commercial surrogacy for children. It is not only for the use of the woman's body that money has exchanged hands, but also for the child itself. As we have seen, a large part of Firestone's objection to the biological family and her desire to diffuse parenting through the structure of the household is her identification of the harmfulness of the idea of "*my* child"; the possessiveness by which some parents relate to the child as an adjunct of themselves, as something over which they have rights of ownership. For Firestone, this parental attitude is encouraged by the ordeal of pregnancy and childbirth, and this is part of her objection to biological reproduction: "'To think of what I went through to have you!'" (208). There is a real danger that surrogacy, while it eliminates the traditional biological connection of mother to child entailed by pregnancy, and in some cases even the genetic connection, might actually intensify the risk of such an attitude, since "what I went through to have you!" also includes a huge and potentially debilitating financial cost.

The objections to surrogacy offered here concern commercial surrogacy, not surrogacy arrangements where a friend or family member agrees to bear a child for someone close to them. Of course, these arrangements are not free of the possibility of going painfully, tragically, wrong. But they are not characterized from the outset by a financial transaction that I have argued commodifies both surrogate and child – whatever the intentions of those involved.

Conclusion

Firestone surely did not envisage that the reproductive technologies she championed in 1970 should have enabled phenomena such as a predatory fertility industry, commercial surrogacy and fertility tourism. Does this demonstrate a naivety on her part about the potential of those technologies to be misused? I think not, although she may have been over-optimistic about the prospects of their *not* being so misused. Firestone was always clear that such technologies *could* be used in oppressive ways. The value of her contribution is that she asks us to think about the conditions that would have to be met for them not to be deployed in ways that entrench exploitation, but that rather unleash their liberatory potential. These might include conditions in which such technologies are available through state-funded healthcare and not through private, profit-making providers; or in which economic structures, employment practices and the social prohibitions against lone parenting do not oblige those who wish to start a family to delay.

Of course, from Firestone's perspective, there remains a problem to the extent that egg-freezing, IVF and surrogacy are all techniques designed to make traditional parenthood possible for those for whom it otherwise would not be. There is a danger that the normalization of such practices would reinforce the hegemony of parenthood and the stigmatization of those who remain childless; and also that it would entrench the idea that to be a parent requires a biological, genetic, relation to a child. This is why Firestone's calls for cultural shifts in our understanding of "parenting" are so important. By this, I mean both her demand that parenting be displaced from its hegemonic position and be considered of equal value alongside other, non-reproductive, lifestyles; and her call for a radical rethinking of what parenting is and in what kind of unit it may occur. I have just used the term "lone parenting." But why should it be considered that to parent outside of the structure of the couple is to parent *alone*? In reality, single parents often share the responsibility and the pleasures of caring for a child with others: both with other members of perhaps an extended family, and with friends who have no genetic relationship to the child. What would it be for these support structures to be recognized as "parenting"? To be given social acknowledgment as ways in which children are raised, and not to be seen as inferior to the nuclear family? Such structures could in fact be given more formal social recognition in order better to materially support them. Why could not parental leave be shared by a nominated person or persons other than a biological or adoptive parent, for example?

Such a rethinking of the unit in which children are raised is what Firestone was trying to achieve with her proposal for 'households.' An extended conception of the reproductive unit could have as just one of its advantages the opening up of the ability to be involved in child-raising for adults who cannot, or do not wish to have biological children of their own. It is possible that under such conditions the very desire for one's *own, genetic,* child, would be diminished; and with it, interest in those technologies that make biological offspring possible for those for whom otherwise it is not.

10

Pregnancy on Trial

"The answer is that there has to be some form of punishment for the woman."
(Donald Trump)

It must not be forgotten that by "reproductive technologies," Firestone meant not only such things as artificial insemination, in vitro fertilization and artificial wombs, but also abortion and contraception. She was by no stretch alone among second wave feminists in calling for the widespread availability of these as a condition of women's emancipation: in fact, this was a call that was widely made by liberal, socialist and radical feminists. But in arguing so forcefully that involuntary reproduction lies at the heart of women's oppression, she presents a particularly powerful account of how contraception and abortion are basic, *necessary* – though not *sufficient* – conditions of gender equality.

In the US, this account is today needed now more than ever. The forces of moral conservatism are rallying for an assault on women's reproductive rights. Trump has allied himself to these forces, promising to nominate pro-life judges to the Supreme Court and to sign into law the Pain-Capable Unborn Child Protection Act, which would enforce a ban on abortion after 20 weeks. He has undertaken to defund Planned Parenthood, which offers sexual health advice to millions of American women. He has said that he will make into permanent law the Hyde Amendment, which is added as a rider to legislation to prohibit taxpayer money from funding abortions, except in cases of rape, incest or threat to the woman's life. And he is committed to dismantling the Affordable Care Act, Obama's signature healthcare plan, which among other things guarantees access to contraception. Pro-choice campaigners fear that Roe v. Wade could be overturned. The wake of Trump's election victory saw women urging each other on social media to obtain long-term forms of contraception such as IUDs, for fear that access to birth control would soon be withdrawn. There is a real danger of a return to the forced pregnancies, forced births and involuntary parenthood that preceded second wave feminism.

And yet, in fact, these things had never really gone away. Access to birth control and abortion in the US (and in other countries) has always been a mixed picture, with the existence of formal, legal rights masking the extent to which a woman's ability to make use of such rights varies drastically with her material circumstances. Without federal provision of reproductive healthcare, in many states it is left to charitable organizations such as Planned Parenthood to provide advice and access to contraception and abortion services. But inevitably in these circumstances, many women without private healthcare are unable to access such services and are therefore denied the means to control their reproductive lives. The threatened assaults on women's reproductive rights today would roll back reproductive freedoms that in fact had never been equally available.

When Trump told MSNBC interviewer Chris Matthews that women who had illegal abortions should face "some form of punishment" he broke an important rule of the pro-life lobby – and quickly recanted. Formally, the pro-life position is that it is the doctors who provide the procedures who should be subject to criminal charges, since they are the ones doing the harm, both to the fetus, and, it is claimed, to the mother. As Anthony Zurcher, a BBC reporter, states, 'The reason for this is simple – to make abortion bans more acceptable to a general public that does not want to see possibly distraught women grappling with unwanted pregnancies sent to prison.'[89] Criticizing Trump's comments, Jeanne Mancini, President of the March for Life

Education and Defense Fund, said: "No pro-lifer would ever want to punish a woman who has chosen abortion."[90]

But what this obscures is that US women are *already* facing punishment for allegedly having chosen abortion. This is taking place in circumstances where pregnancy complications, miscarriage or stillbirth are being interpreted as being *suspicious*; as suggesting that a woman has acted to induce her own abortion. And it is in turn part of a creeping criminalization of pregnancy that sees pregnant women increasingly subjected to surveillance and investigation, especially if they are on low incomes, are women of color or from immigrant backgrounds.

When birth control and abortion exist as formal rights but in reality are denied to many people, one result is that women are forced into seeking illicit ways of controlling their reproductive biologies. This will only happen more frequently as the result of measures to close down still further women's access to reproductive health services: not only the Hyde Amendment; but also attempts at state level to introduce versions of the federal Religious Freedom Restoration Act, that would allow medical professionals to refuse contraception and abortion on grounds of conscience; and the Hobby Lobby Supreme Court judgment, which established that some companies can opt out of including birth control coverage in their employee health insurance packages. At the same time, there is a burgeoning of laws that are being used specifically to do what the pro-life lobby denies is its agenda: to punish women for having – or appearing to have had – illegal abortions. In reality, anti-abortion forces are producing the very phenomenon that they then seek, through the back door, mercilessly to punish.

Feticide and "personhood"

In July 2015, Purvi Patel, an Indian American woman, was sentenced by an Indiana court to 20 years' imprisonment for feticide and child neglect. The prosecutor had requested 40 years, and Patel in fact received 30, but with 10 years suspended. In July 2013, Patel had given birth to a fetus[91] in the bathroom of her family home. She claims that the fetus was stillborn. She placed its remains in a bag of trash and left it in a dumpster behind her family's restaurant, before heading to a medical facility to seek treatment for her uninterrupted bleeding. While there, doctors and then a police officer questioned her, and Patel admitted to having recently given birth. A search for the baby – whom doctors believed might possibly still be alive – ended in the discovery of its remains.

Patel's conviction and imprisonment rightly caused outrage among women's reproductive rights advocates. In order to punish her for what it was claimed was an illegal, self-induced abortion, Indiana's prosecutors drew upon the state's feticide law to charge her for killing her fetus. There are so many things wrong with this that it is hard to know where to start. Firstly, the claim that Patel had brought about an abortion was based on some mobile phone text messages, which seemed to suggest she had illegally bought abortion-inducing drugs online from Hong Kong. Patel denies that she took the drugs, and no traces of them were found in her body. The drugs themselves are legally available in Indiana for the purpose of inducing abortion, but they must be obtained through prescription. Secondly, however, even had Patel taken the drugs, she would only have been bringing about illicitly a termination to which she was, probably, legally entitled. As per Roe v. Wade, abortion is legally available in Indiana up until the point of viability, which is taken to be at around the end of the second trimester. Patel's legal team argued at her trial that the fetus was stillborn at 24 or 25 weeks, although Patel says that she had believed the pregnancy was less advanced than this. It is not hard to conceive of some of the reasons why Patel might have considered ending her pregnancy through illicit means. Indiana makes it very difficult for a woman to obtain an abortion. There are just 12 abortion providers in the entire state, and women must go through a process of two separate visits, counseling and a waiting period to obtain a termination. Patel was acting as the carer to her parents and grandparents with whom she lived: highly religious and morally conservative, her family does not believe in sex before marriage, and she had kept her pregnancy secret from them.

Most worryingly, however, is that in order to punish her for her alleged abortion the prosecutors charged her under Indiana's feticide law. Feticide laws, which many states now have, supposedly protect women and their unborn children from the violence of third parties, such as abusive partners. At the same time, the prosecutors charged her with child neglect, alleging (on the basis of a discredited "lung float test") that the fetus had taken at least one breath upon its delivery. The prosecution's pathologist testified that the fetus was in fact between 25 and 30 weeks old, and therefore might have lived, had Patel taken it for immediate medical attention. Her prosecutors were therefore in the paradoxical position of accusing Patel of a crime which required that her unborn fetus had been killed – feticide – and of a crime that required that it had been born alive – child neglect. They attempted to resolve this by arguing that a feticide charge could be made for attempting to end a pregnancy even if the fetus survived.

Patel's imprisonment ended in September 2016, after the Indiana Court of Appeal had reconsidered her case in July. The appeal judge overturned the feticide verdict on the basis that the use of the feticide law to punish women for self-inducing abortions was an "abrupt departure" from the intent of the law as shown by previous usage. Ruling that the prosecution had proved that the child was born alive, but that they had not proved it would have survived had it received medical attention, she also reclassified the child neglect charge to the lesser one of felony neglect, ordering that Patel be resentenced. Patel was eventually given a sentence of 18 months, which was less time than she had already served. For many who have followed her case, Patel still has not received justice; there is, after all, no conclusive evidence that she did not merely suffer the accidental miscarriage of a stillborn fetus, and respond in understandable trauma.

And Patel's conviction is not a one-off, an anomaly or an outlier. Rather, it is part of a strategy to establish feticide laws as a way of punishing women who are deemed to have had illegal abortions. It is part of a movement to establish the legal "personhood" of fetuses, embryos and even fertilized eggs, which itself is a route towards the banning of abortion altogether. Indeed, this would even threaten some forms of contraception, such as some IUDs, which prevent a fertilized embryo from implanting in the womb's lining, and the morning-after pill. At the time of Patel's conviction, attorney Katherine Jack commented that it "basically sets a precedent that anything a pregnant woman does that could be interpreted as an attempt to terminate her pregnancy could result in criminal liability."[92] Also responding to the issue of precedent, Sue Ellen Braunlin, co-president of the Indiana Religious Coalition for Reproductive Justice, said, "I know how aggressively prosecutors in this state were going to expand the application of this feticide law. This is what they wanted so badly."[93]

Criminalizing miscarriage, criminalizing despair

In fact, Patel was not the first Indiana woman to be charged under its feticide laws. In 2012, Bei Bei Shuai, an immigrant to the US from China, was charged with the attempted feticide and the murder of her fetus. She faced a prison sentence of 45 years to life. In December 2010, and after being abandoned by her boyfriend, Shuai had tried to end her own life by consuming rat poison. She survived, but just over a week later and while still in hospital, the fetus, at 33 weeks gestation, was showing signs of distress and was delivered by Caesarian section. A few days later, doctors removed the infant from life support, and allowed it to die in Shuai's arms. Shuai 'begged for her own life to be taken so that her child's might be spared.'[94] The murder charge was eventually dropped in 2013, after Shuai pleaded guilty to a charge of criminal recklessness. She had served 1 year in jail.

It seems incomprehensible that a woman so deeply despairing as to attempt suicide, and so grief-stricken at the loss of her baby, should be thrown in jail rather than cared for. But it is also entirely in keeping with what Lynn Paltrow, executive director of the National Advocates for Pregnant Women (NAPW), notes is a practice of "making pregnant women – from the time an egg is fertilized – subject to state surveillance, control and extreme punishment."[95] In Texas in 2003, for example, a local district attorney took advantage of a new feticide law to write to local doctors, requiring them to report pregnant women who were drug users; more than 50 women were reported and charged with crimes. In a 2014 opinion piece in *The New York Times*, Paltrow and her co-author, Fordham University professor, Jeanne Flavin, report on the increasing use of laws to arrest women who have no intention of ending their pregnancies and to force unwanted treatments upon them. Their examples include a critically ill woman who was forced by a judge to undergo a Caesarian section at 26 weeks, although he knew it might kill her – neither she nor the fetus survived; an Iowa woman who fell down the stairs and, after arriving at hospital to check that her unborn baby was unharmed, was arrested for attempted feticide; a Louisiana woman, hospitalized for vaginal bleeding, who was incarcerated for a year on second degree murder charges, before it was revealed that she had suffered a miscarriage. The NAPW say that they have documented hundreds of cases in which women have faced criminal charges for "suspicious miscarriages," for drug use or for trying to induce their own terminations.

One thing these legal measures do is to establish beyond doubt the status of women as mere incubators for something that is more important than them. The alleged interests of the fetus are deemed to legislate the overriding of a woman's most basic of human rights: to physical liberty, and to the ability to refuse medical procedures that may harm or kill them.

Another thing they do is to criminalize uncertainty. Both Patel and the Iowa woman mentioned by Paltrow, Christine Taylor, had expressed doubts about whether they wished to continue with their pregnancies. Their stories and those of other women show that any articulation of ambivalence about being pregnant may be a dangerous thing, for it establishes a context in which subsequent complications, miscarriage or stillbirth, are liable to a criminalizing interpretation.

And yet another is to place still further beyond the reach of many women access to reliable reproductive information and safe treatment, for fear that should they approach medical practitioners they may be subject to investigation and punishment.

Conclusion

It is women of color who "are bearing the brunt of [these] unscientific laws and misplaced moral outrage against abortion," as Yamani Hernandez, Director of the National Network of Abortion Funds, notes.[96] And this is a further instance of a pattern of racialized reproductive injustice that is evident in other areas too, as grassroots organizations such as SisterSong point out.[97] Shockingly, maternal mortality rates are actually rising in the US (the only developed nation of which this is true) and black women are dying at four times the rate of white women, at all income levels.[98] The reasons for this are complex, but are thought to include the stresses of enduring racism on maternal and fetal health, and a pattern of many black women delaying or not initiating prenatal care. The growing criminalization of pregnancy will cost still further lives.

This is not to suggest that the US is the only country in which advances in reproductive freedoms gained through second wave feminism are at risk. In the UK there have been moves (so far unsuccessful) to reduce the time limit within which abortions may take place. Trade unions in Northern Ireland have warned that women who present at hospitals with miscarriages may face abortion questioning, after a woman was found guilty of procuring abortion pills in 2016.[99] The rise of right-wing populism across Europe threatens women's reproductive rights and health, as evidenced in the recent attempts in Poland to introduce a total abortion ban, defeated only when tens of thousands of women boycotted work and took to the streets in protest. In Nicaragua, a total abortion ban was introduced in 2008, with women and their physicians facing jail for terminations, even in the instances of rape, incest and threat to the woman's life. Child victims of rape are forced to give birth. Doctors and nurses are fearful of providing pregnant women with even life-saving medical care in relation to cancer, cardiac emergencies and HIV/AIDS, for fear of an ensuing miscarriage. The situation, according to a recent Amnesty International visitor, is one of "sheer horror."[100]

But the US is leading the way in the West in a rolling back of hard-won reproductive rights, both through the use of legislation to withdraw from women the material supports they need to make use of those rights, and through laws that make of pregnancy an object of surveillance, discipline and punishment.

Firestone was right. US pregnancy *is* barbaric. And unless there is an effective campaign of feminist resistance, it is set to become still more so.

11

Conclusion

Firestone's is still the only historical-material explanation of the ubiquity of women's oppression, arguing that in the circumstances of early humans, the biological facts and consequences of reproductive sex were sufficient to establish societies in which men predominated in the majority of social functions. She may be wrong to infer from this that it remains the case that if male domination is to change, then biology has to be changed: that transforming reproductive biology through technology is a *necessary* condition of social change.

But even if she is wrong in this, that detracts neither from her historical-material explanation nor from the need to examine the continuing role of reproduction in contributing to women's oppression. Changing biology may not be necessary: it certainly isn't *sufficient*. But in explaining, and thus demystifying, the ubiquity of patriarchy, she utterly *denatures* it. Removing that misunderstanding is already an achievement.

In fact, I think Firestone *is* wrong that there needs to be a transformation of biology. For me, key to her contribution to feminism is her materialism, which insists on the fundamental inseparability of production and reproduction. These are two sides of the same coin, she insists; indivisible aspects of the economic problem always facing human beings – of how to create and sustain life.

But does "material change" have to mean altering biology? I think not. For another aspect of her contribution is her imaginative daring; her refusal to be bound by "capitalist realism": the belief that there is no alternative to the endless instrumentalization of others and ourselves. The feminist thinker too often recalled only as the one who wanted artificial wombs is also, as we have seen, committed to a socialist transformation of production and distribution, and not as an adjunct to her program for women's liberation but as a necessary condition of it. She sees clearly that there can be no freeing of women from their domestic roles for as long as the responsibility for maintaining life (children, and we might add, caring for the elderly, or the unwell) is privatized and put onto the individual or the family. Her analysis entails that redress of gender and other inequalities does not lie within the hands of individuals alone; that it cannot be a simple matter of individuals thinking differently about things, deciding to act or feel otherwise than they do, or adjusting themselves differently to the world that confronts them. The fundamentals of that world must be changed instead.

For me, what her analysis points to is that reproduction must be *voluntary*: for which is required access to contraception and abortion, and a cultural shift in our attitudes to people who are not parents. That it must be *materially supported*: which requires a radical rethinking of how childrearing is paid for, and a shift from the privatization of these costs to their collectivization, through such things as increased public/state support of childcare provision and parental employment leave, and the provision of an income upon which those who need or wish to concentrate full-time on parenting can live and raise their children with dignity. And it requires that reproductive possibilities be *extended beyond* the nuclear family through the support of alternative structures for the raising of children.

We need today more than ever a *feminist* radical politics that will put these issues on the agenda. The challenge that Firestone poses to us today is how to accomplish the changes that her analysis calls for. This is the provocation of her book: "What then, is to be done?" If not Firestone's own (tentative) proposals, what then instead?

12

Epilogue: Airless Spaces

Firestone's second, and final, book, appeared 28 years after *The Dialectic of Sex*, in 1998. Shortly after publication of the *Dialectic*, Firestone had withdrawn from feminist activism and had ceased to publish. One of the most dynamic figures of the women's liberation movement had effectively vanished from it, just as the decade of second wave feminism was really getting going. The precise reasons for Firestone's withdrawal are unknown, although her second book gives us glimpses into what her life had been since 1970.

Airless Spaces consists of 51 vignettes, ranging in length from a short paragraph to a few pages, and grouped under the headings 'Hospital,' 'Post-Hospital,' 'Losers,' 'Obits' and 'Suicides I Have Known.' These are centered upon the lives of briefly introduced persons living in New York State, and sometimes within or around its mental institutions. There is nothing to identify the sketches as being either fictional or drawn from life, and no explicit identification of the 'I' that appears after the first few stories as Firestone herself. But the back cover tells us that 'Refusing a career as a professional feminist, Shulamith Firestone found herself in an "airless space" – approximately since the publication of her first book *The Dialectic of Sex*.'

What might be meant by an 'airless space' is suggested in the brief 'Frontispiece' to the book. Firestone (as I shall identify the narratorial voice) tells us of a dream in which she is on a sinking ship, a 'luxury liner like the Titanic.' While on the upper decks the doomed people try to disguise their fear with 'gaiety and mirth,' Firestone goes down into the ship, searching for a space that would 'supply an air pocket,' in which she might survive 'even after the boat was fully submerged until it should be found.'

Eventually she stows herself inside a refrigerator. Waking in 'panic' and convinced 'the disaster was real,' Firestone calls the authorities. There is indeed a liner sinking, she is told, but it is in the Bermuda Triangle, so no rescue attempt will be made.

The people in the stories are, like the dreamer of the Frontispiece, the ones who cannot stay on the upper decks. Isolated by their inability to feign merriment, they must search alone for some space that, however constricted, allows them to continue holding on. For insomniac Bettina, deprived of even the smallest 'buffer' of sleep by the hospital's 'high, narrow' bed, its 'fifteen-minute head counts all night long,' her medication and the warmth of her dormitory, this is found for the few brief nights in which, having secreted away the thermostat control, she is able to lie in chilled air, 'still as a corpse,' in simulation of sleep (32–3). For the elderly Pauline, once a 'noted pianist' and now a 'paranoid' whom only a wealthy background protects from 'permanent institutionalization,' it is the classical music played on radio station WISS; until the 'weak works' and the advertisements so predominate over the music she loves that she turns the station off for good (100, 102). Lynn, who lives on disability checks, dates Neil, albeit just as friends, until she can no longer bear his self-absorption and obvious contempt for her.

The protagonists of these stories are struggling to maintain some degree of happiness, dignity or personal autonomy in a world in which the shattering of hopes for intimacy or meaning is the norm. Hospitalization, the sketches suggest, is rarely a form of rescue. Instead, in-patient life produces routine humiliations that dehumanize and depersonalize, producing as much as alleviating the conditions that the professionals purport to diagnose. Corrine refuses to shower until the hospital water quality has been improved; it is only when a forced shower, conducted with 'brutal...merriment,' leaves her hair 'matted and stiff with chemicals' from unwashed-out soap that she 'began to look like a mental patient' (14–16). Ellis Martin Sheen lives on food stamps and participates as little as possible in his day-group meetings, until one day he demands irately to know who has ever seen him drinking. His social worker, 'needing a handy pigeon-hole to account for Martin's emotional illness,' has 'decided that Martin was an alcoholic' (81). Leon Feldsher is registered disabled 'by virtue of mental distress...though no-one was quite sure just which psychiatric ailment he had' (82).

As the book progresses, it becomes increasingly apparently autobiographical, increasingly clear that the 'I' of these tales at least is the author of the famous feminist manifesto. We learn of Firestone's visit to Valerie Solanas, the author of the radical feminist SCUM Manifesto who shot Andy Warhol and art critic Mario Amaya in 1968. 'It was not that I admired her deed,' Firestone tells us. 'Nor did I particularly value her book,' which she had never considered 'serious feminist theory' (130). Valerie in turn takes a 'scored...all over' copy of the *Dialectic* from her shelf and tells her, 'I didn't like your book,' though Firestone is unperturbed (131). 'Some time later,' Firestone tells us, she often sees Valerie on the streets, evidently seriously unwell and begging for quarters (131–2). Firestone seemingly reverts to the third person to write about herself from a distance in a number of stories. In 'Myrna Glickman,' for example, we hear about 'Rozzie,' a 'founding member' of the Stanton-Anthony Brigade of the New York Radical Feminists, who is betrayed by her friend, the straitlaced but easily led Myrna (117). 'Emotional Paralysis' describes the disabling inertia of life after release from mental hospital. Its unidentified 'she' 'sometimes recognized on the faces of others joy and ambition and other emotions she could recall having had once, long ago' (59). But she herself cannot read, or write, the 'old excitement of creation' being gone and in its place an inability to 'care about anything' (59).

The volume's final, and longest, piece, 'Danny,' tells the devastating story of Firestone's brother: of their initial closeness and mutual support in the face of warring parents, 'heavy-duty religious observation...imposed on us' (151) and neighborhood anti-semitism; of their growing estrangement as Danny learns to hate girls, and Firestone becomes secularized; and of Danny's apparent suicide in mysterious circumstances. The uncertainty around his death, Firestone tells us in the final sentence of the book, 'contributed to my own growing madness – which led to my hospitalization, medication, and a shattering nervous breakdown' (160).

* * *

There is a danger in ending a book on Firestone in this way. It is the danger of encouraging a mistake that occasionally even my very perceptive students make.

The mistake is to think that Firestone's history of acute psychological distress somehow explains the *Dialectic*, allowing us to see that the meaning of its radicalism, its stridently non-conformist worldview, was always incipient mental illness. The *Dialectic* thus becomes read as a *symptom* of Firestone's "madness." Which means, of course, *not* reading it. *Not* engaging with its ideas; but instead, dismissing it from the scene of serious political and theoretical engagement.

But this is to get things the wrong way round. We must not use "mental illness" to depoliticize radical theory; but use radical theory to politicize "mental illness." The urgent task is to identify and analyze the social and economic structures that work to produce a widespread psychological distress, to which are attributed diagnostic labels.

In a sense, Firestone had always been writing about airless spaces. What else is the patriarchal nuclear family, in her analysis, than a place in which one looks for shelter, only to discover that one cannot breathe? Despite the vast differences between the two books, I therefore propose reading *Airless Spaces* as a kind of coda to *The Dialectic of Sex*. In its documenting of a brutal and alienating society in which people are subordinated to profit, it depicts a nightmare inversion of the more humane society glimpsed in the *Dialectic*. Read together, the two books proclaim that we don't own each other: that we are equals: that we are all vulnerable and in need of care. They constitute an exhortation to mobilize our energies in the fight against the structures of twenty-first-century patriarchal capitalism that prevent us from seeing this and from acting accordingly.

Endnotes

1. Marge Piercy, *Woman on the Edge of Time* (London: The Women's Press, 1979), pp.105–6.
2. Piercy, p.102.
3. Nina Power, *One Dimensional Woman* (Winchester and Washington: Zer0 Books, 2009); Laurie Penny, *Meat Market: Female Flesh under Capitalism* (Winchester and Washington: Zer0 Books, 2010). See also Maureen Nappi, 'Shulamith Firestone: Cybernetics and Back to a Feminist Future,' in *Situations*, 6:1&2, pp.187–212.
4. Mandy Merck and Stella Sandford, eds, *Further Adventures of The Dialectic of Sex: Critical Essays on Shulamith Firestone* (New York: Palgrave Macmillan, 2010).
5. Ann Snitow, quoted by Merck in 'Shulamith Firestone and Sexual Difference,' in *Further Adventures*, p.13.
6. Alice Echols, *Daring to Be Bad: Radical Feminism in America, 1967–1975* (Minneapolis: University of Minnesota Press, 1989), p.67.
7. Susan Faludi, 'Death of a revolutionary,' *The New Yorker*, 15 April 2013, http://www.newyorker.com/magazine/2013/04/15/death-of-a-revolutionary
8. Faludi.
9. Alison Jaggar, *Feminist Politics and Human Nature* (Brighton: Harvester Press, 1983), p.93.
10. Echols.
11. Mary O'Brien, *The Politics of Reproduction* (Boston, London and Henley: Routledge and Kegan Paul, 1981), p.79.
12. See Sam McBean, *Feminism's Queer Temporalities* (Abingdon and New York: Routledge, 2016), p.55.
13. See for example Nancy Fraser, 'Feminism, Capitalism and the Cunning of History,' *New Left Review*, 56, 2009; also Power.

14. Jo Freeman, 'On Shulamith Firestone,' https://nplusonemag.com/issue-15/in-memoriam/on-shulamith-firestone/
15. Merck, p.11. In the years subsequent to the *Dialectic*'s publication, when Firestone had retired from public life and requested that the documentary not be made public, it was recreated on a shot-by-shot basis by feminist filmmaker Elizabeth Subrin, who substitutes for the male filmmakers' voiceover her own narration. The remade film apparently elicited from the older Firestone responses ranging from cautious acceptance to anger and dismay.
16. The following account is taken from Echols, p.117.
17. Echols, p.142.
18. Echols, p.140. The quotation is from a Redstockings flyer from 1973.
19. Echols, p.196.
20. Echols, p.152, p.192 and elsewhere.
21. Betty Friedan, *The Feminine Mystique* (London: Penguin, 2010).
22. Echols, p.141.
23. NARAL: Pro-Choice America, 'The safety of legal abortion and the hazards of illegal abortion,' 1 January 2015, p.5, http://www.prochoiceamerica.org/media/fact-sheets/abortion-distorting-science-safety-legal-abortion.pdf
24. Echols, p.141.
25. Echols, p.141.
26. See the memoir by Linnea Johnson, 'Something real: Jane and me. Memories and exhortations of a feminist ex-abortionist,' http://www.cwluherstory.org/something-real-jane-and-me-memories-and-exhortations-of-a-feminist-ex-abortionist.html
27. Sarah Franklin, 'Revisiting Reprotech: Firestone and the Question of Technology,' in *Further Adventures*, p.45.
28. Stella Sandford, 'The Dialectic of the Dialectic of Sex,' in *Further Adventures*, p.235.

29. See for example, O'Brien, for whom Firestone confuses mere dichotomy for dialectic (in the former, opposites can coexist; in the latter, opposition commands mediation): O'Brien, p.80. For both Stella Sandford and Tim Fiskin, however, there are several different dialectics being pointed to, at least implicitly, in Firestone's text. See Sandford, pp.241–3; and Tim Fiskin, 'Technology, Nature, and Liberation: Shulamith Firestone's Dialectical Theory of Agency,' in *Further Adventures*, p.199. Alison Assiter has argued that the notion of 'sexual class' doesn't work, since there can be no correlatives for Marx's modes and forces of production, or surplus value (pp.72–3). Caroline Bassett, however, has pointed out that it is often unclear quite how Firestone is using Marx: at times, she observes, Marxism seems to be less an 'operational model' and more an 'allegorical model' for an actually quite different kind of revolution. See Bassett, 'Impossible, Admirable, Androgyne: Firestone, Technology and Utopia,' in *Further Adventures*, p.91.
30. Quoted in Juliet Mitchell, 'Women: The Longest Revolution,' *New Left Review*, 40, December 1966, https://www.marxists.org/subject/women/authors/mitchell-juliet/longest-revolution.htm
31. Friedrich Engels, 'Origins of the Family, Private Property, and the State,' https://www.marxists.org/archive/marx/works/1884/origin-family/ch02c.htm
32. For different articulations of this criticism, see Michelle Barrett, *Women's Oppression Today: The Marxist/Feminist Encounter* (London and New York: Verso, 1988); Alison Assiter, *Althusser and Feminism* (London: Pluto, 1990); Linda J. Nicholson, *Gender and History: The Limits of Social Theory in the Age of the Family* (New York: Columbia University Press, 1986); Sarah Walby, *Theorizing Patriarchy* (Oxford: Basil Blackwell, 1990).
33. See Firestone, p.10. I am in agreement here with Fiskin's

interpretation of Firestone on the relationship of the nuclear to the biological family.

34. Donna Haraway, *Simians, Cyborgs, and Women: The Reinvention of Nature* (New York: Routledge, 1991), p.10.
35. Barrett, p.12.
36. Simone de Beauvoir, *The Second Sex*, trans. H. M. Parshley (London: Vintage, 1997), p.295. All subsequent references are to this edition.
37. Jane Alpert, *Mother Right: A New Feminist Theory* (Pittsburgh: Know, Inc., 1974), pp.7–9, http://cdm15957.contentdm.oclc.org/cdm/ref/collection/p15957coll6/id/669
38. Hélène Cixous, *The Newly Born Woman* (London: I.B. Tauris, 1996), p.90.
39. Shulamith Firestone, *Airless Spaces* (South Pasadena: Semiotext(e), 1998), p.130.
40. Elizabeth V. Spelman, *Inessential Woman: Problems of Exclusion in Feminist Thought* (Boston: Beacon Press, 1988), pp.126–9.
41. Rick Jervis, 'Texas's maternal death rates top most industrialized countries,' *USA Today*, 10 September 2016, http://www.usatoday.com/story/news/health/2016/09/10/texas-maternal-mortality-rate/90115960/
42. World Health Organization website, http://www.who.int/mediacentre/factsheets/fs348/en/
43. Juliet Mitchell, *Psychoanalysis and Feminism* (London: Allen Lane, 1974), pp.347–50; Rosalind Delmar, 'Introduction' to Shulamith Firestone, *The Dialectic of Sex: The Case for Feminist Revolution* (London: The Women's Press, 1979), p.9.
44. Penny, p.53.
45. Hortense Spillers, *Black, White and in Color* (Chicago: University of Chicago Press, 2003), pp.159–64.
46. Angela Y. Davis, *Women, Race and Class* (London: The Women's Press, 1982), p.181–2.
47. See Davis, pp.70–86. See also bell hooks, *Ain't I a Woman:*

Black Women and Feminism (New York: Routledge, 2015), pp.124–31. Davis quotes Elizabeth Cady Stanton in a letter to the *New York Standard* in 1865, thus: '"as the celestial gate to civil rights is slowly moving on its hinges, it becomes a serious question whether we had better stand aside and see 'Sambo' walk into the kingdom first"' (70). Stanton and Susan B. Anthony had campaigned against slavery, but eventually argued successfully for the dissolution of the Equal Rights Association, an alliance of white feminists and black liberationists. Firestone nonetheless uncritically groups Stanton, Anthony and black civil rights activist Sojourner Truth together as the true radicals of the first wave feminism, and named her cell of the New York Radical Feminists the 'Stanton-Anthony Brigade.'

48. bell hooks, *Ain't I*, especially pp.136–58.
49. bell hooks, *Yearning: Race, Gender, and Cultural Politics* (New York and Abingdon: Routledge, 2015), pp.41–50.
50. Carolyn Steedman, *Landscape for a Good Woman: A Story of Two Lives* (London: Virago, 1986), see esp. pp.48–51, 72–82.
51. This practice was revealed by the 1974 court case of Relf v. Weinberger. See the Southern Poverty Law Center website, https://www.splcenter.org/seeking-justice/case-docket/relf-v-weinberger
52. O'Brien, p.79.
53. Haraway, pp.9–10.
54. Franklin, pp.31–2, and 45. See also, Walby, p.67; Bassett; Susanna Paasonen, 'From Cybernation to Feminization: Firestone and Cyberfeminism,' in *Further Adventures*.
55. Mark Fisher, *Capitalist Realism: Is There No Alternative?* (Winchester and Washington: Zer0 Books, 2009), p.33.
56. Stevi Jackson, 'Questioning the Foundation of Heterosexual Families,' in *Further Adventures*. The Segal quotation is from: Lynne Segal, *Is the Future Female?* (London: Virago, 1987), pp.5–6.

57. Power, p.65–6.
58. For example, see Alex Tabarrok for an argument against the 'Ludite Fallacy – the idea that new technology destroys jobs,' in 'Productivity and Unemployment,' Marginal Revolution, http://marginalrevolution.com/marginalrevolution/2003/12/productivity_an.html. For an opposing view, see Paul Krugman, 'Sympathy for the Luddites,' *The New York Times*, 13 June 2013, http://www.nytimes.com/2013/06/14/opinion/krugman-sympathy-for-the-luddites.html?_r=0. The argument that in the future machines will maintain machines is from Marshall Brain, *Robotic Nation and Robotic Freedom* (BYG Publishing, 2013).
59. See for example, Marshall Brain, for an argument that simultaneously mounts a critique of capitalism's wealth-concentrating operations and calls for a UBI as a way of 'Turbo-charging' capitalism.
60. Heather Stewart, 'John McDonnell: Labour taking a close look at universal basic income,' *The Guardian*, 5 June 2016, http://www.theguardian.com/politics/2016/jun/05/john-mcdonnell-labour-universal-basic-income-welfare-benefits-compass-report
61. In a pilot project in Otjivero-Omitara, Namibia, in 2008–9, it was found that a Basic Income Grant (BIG) actually increased work effort and economic activity. See Claudia and Dirk Haarmann, Basic Income Grant Coalition website, http://www.bignam.org/BIG_pilot.html. In Canada in the 1970s, a Guaranteed Annual Income (GAI) experiment conducted among the low income population of Dauphin, Manitoba, showed that there was only a small reduction in work effort, and that this was restricted to new mothers (who elected to stay home with their newborns rather than go out to work) and to teenagers (who were relieved of pressure to support their families). See Vivian Belik, 'A town without poverty,' *The Dominion*, 5 September 2011, http://

www.dominionpaper.ca/articles/4100. See also Derek Hum and Wayne Simpson, 'A Guaranteed Annual Income? From Mincome to the Millennium,' *Policy Options*, 79, January–February 2001, http://archive.irpp.org/po/archive/jan01/hum.pdf.

62. See Jaggar, p.154; and Jackson, p.114.
63. Ariel Levy, *Female Chauvinist Pigs: Women and the Rise of Raunch Culture* (New York: Simon and Schuster, 2005); Natasha Walter, *Living Dolls: The Return of Sexism* (London: Virago, 2010).
64. See Michel Foucault, *The History of Sexuality: The Will to Knowledge, Vol. 1*, trans. Robert Hurley (London: Penguin, 1998).
65. Power, pp.144–8.
66. Power, p.150.
67. Sarah Marsh and *Guardian* readers, 'The gender fluid generation,' *The Guardian*, 23 March 2016, https://www.theguardian.com/commentisfree/2016/mar/23/genderfluid-generation-young-people-male-female-trans68. Delmar, p.5.
69. Spelman, p.130.
70. Gillian Howie, 'Sexing the State of Nature: Firestone's Materialist Manifesto,' in *Further Adventures*.
71. Mia Fahlén and Gertrud Åström, 'Women's bodies aren't simply containers,' *The Local*, 5 March 2015, http://www.thelocal.se/20130305/46492
72. Victoria Ward, 'Louise Brown, the first IVF baby, reveals family was bombarded with hate mail,' *The Telegraph*, 23 November 2016, http://www.telegraph.co.uk/news/health/11760004/Louise-Brown-the-first-IVF-baby-reveals-family-was-bombarded-with-hate-mail.html
73. Penny, p.46.
74. Toby Helm and Rowena Mason, 'Leadsom told to apologise,' *The Guardian*, 9 July 2016, https://www.theguardian.com/politics/2016/jul/09/andrea-leadsom-told-to-apologise

75. Jessica Valenti, *Why Have Kids: A New Mom Explores the Truth about Parenting and Happiness* (Las Vegas: Amazon Publishing, 2012).
76. Eva Wiseman, 'We need to talk about egg-freezing,' *The Guardian*, 7 February 2016, https://www.theguardian.com/society/2016/feb/07/life-on-hold-with-frozen-eggs
77. See Franklin, pp.47–8 on IVF and its intensification of women's reproduction; and also for an introduction to some of the feminist literature on IVF.
78. HFEA website, http://www.hfea.gov.uk/46.html
79. EggBanxx website, https://www.eggbanxx.com/egg-freezing-costs
80. Sarah Marsh and *Guardian* readers, 'Are you worried about your fertility? Young people share their stories,' *The Guardian*, 20 April 2016, https://www.theguardian.com/commentisfree/2016/apr/20/are-you-worried-about-your-fertility-young-people-share-their-stories
81. Viv Groskop, '"Social" egg-freezing is a hideous fertility gamble,' *The Guardian*, 9 February 2016, https://www.theguardian.com/commentisfree/2016/feb/09/social-egg-freezing-fertility-infertility-parents-children
82. The term "pre-pregnant" was coined in 2006 by the *Washington Post* in response to the Centers for Disease Control and Prevention issuing advice to all women of childbearing age to protect their pre-conception health through behavior and diet, regardless of whether or not they planned to become pregnant. For critical perspectives on this, see Valenti, *Why Have Kids*, and Georgetown University professor Rebecca Kukla (quoted in Valenti, and in Amy Williams, 'Warning: you could be pre-pregnant,' *Ms Magazine* blog, 26 January 2011, http://msmagazine.com/blog/2011/01/26/warning-you-could-be-pre-pregnant/)
83. 'Expensive IVF add-ons "not evidence-based,"' NHS website, 28 November 2016, http://www.nhs.uk/

news/2016/11November/Pages/Expensive-IVF-add-ons-not-evidence-based.aspx

84. See for example, comments by fertility specialists Professor Robert Winston: https://www.theguardian.com/science/2007/may/31/medicineandhealth.health; and Professor Charles Rodeck: https://www.theguardian.com/society/2007/jul/15/health.medicineandhealth
85. Michael Safi, 'Baby Gammy's twin can stay,' *The Guardian*, 14 April 2016, https://www.theguardian.com/lifeandstyle/2016/apr/14/baby-gammys-twin-sister-stays-with-western-australian-couple-court-orders
86. Suzanne Moore, 'The case of Baby Gammy,' *The Guardian*, 4 August 2016, https://www.theguardian.com/commentisfree/2014/aug/04/baby-gammy-thailand-surrogacy-repulsive-trade-pattaramon-chanbua
87. Divya Gupta, 'Inside India's surrogacy industry,' *The Guardian*, 6 December 2011, https://www.theguardian.com/world/2011/dec/06/surrogate-mothers-india
88. I am taking the idea of 'moral climate' from my colleague, Bob Brecher. Brecher suggests that the institutionalization of surrogacy could constitute a 'morality-affecting harm': in other words, although it is conceivable that none of the individuals directly involved in a particular surrogacy agreement is harmed by it, that the agreement nonetheless causes harm by reinforcing wider social acceptance of 'people's making use of each other,' and particularly of women's treatment in terms of commodity. Bob Brecher, 'Surrogacy, Liberal Individualism and the Moral Climate,' in *Moral Philosophy and Contemporary Problems*, ed. J. D. G. Evans (Cambridge: Cambridge University Press, 1987), p.195.
89. 'Trump abortion row: Republican front-runner changes stance,' BBC News website, http://www.bbc.com/news/world-us-canada-35931103

90. 'Trump abortion row: Republican front-runner changes stance,' BBC News website, http://www.bbc.com/news/world-us-canada-35931103
91. The choice of which words to use in discussing these situations is difficult, since there is no neutral language. "Embryo," "fetus," "baby" and "child," all incline toward one moral and political view over another. Likewise for "woman" or "mother." I will use "fetus" in relation to a pregnancy beyond 11 weeks and until 42 weeks, except where the sense of the sentence requires 'baby' or 'child' since that is what is being claimed.
92. Jessica Glenza, 'Purvi Patel case,' *The Guardian*, 2 April 2015, https://www.theguardian.com/us-news/2015/apr/02/purvi-patel-case-alter-reproductive-rights-indiana
93. Jessica Glenza, 'Purvi Patel case,' *The Guardian*, 2 April 2015, https://www.theguardian.com/us-news/2015/apr/02/purvi-patel-case-alter-reproductive-rights-indiana
94. Ed Pilkington, 'Indiana prosecuting Chinese woman,' *The Guardian*, 30 May 2012, https://www.theguardian.com/world/2012/may/30/indiana-prosecuting-chinese-woman-suicide-foetus
95. Jessica Valenti, 'It isn't justice for Purvi Patel,' *The Guardian*, 2 April 2015, https://www.theguardian.com/commentisfree/2015/apr/02/it-isnt-justice-for-purvi-patel-to-serve-20-years-in-prison-for-an-abortion
96. Molly Redden, 'Purvi Patel has 20-year sentence reduced,' *The Guardian*, 22 July 2016, https://www.theguardian.com/us-news/2016/jul/22/purvi-patel-abortion-sentence-reduced
97. Civil Liberties and Public Policy Hampshire website, https://clpp.hampshire.edu/leadership-programs/rrasc/host-sites/sistersong-women-color-reproductive-justice-collective
98. Our Bodies Ourselves website, http://www.ourbodiesourselves.org/2016/03/increased-numbers-of-black-women-

dying-during-pregnancy-and-childbirth/
99. Henry McDonald, 'Northern Irish women who miscarry,' *The Guardian*, 13 April 2016, https://www.theguardian.com/world/2016/apr/13/northern-ireland-women-miscarry-abortion-questioning-unite
100. CNN website, 'Nicaragua abortion ban "cruel and inhuman disgrace,"' http://edition.cnn.com/2009/HEALTH/07/28/nicaragua.abortion.ban

Select Index

Zero Books

CULTURE, SOCIETY & POLITICS

Contemporary culture has eliminated the concept and public figure of the intellectual. A cretinous anti-intellectualism presides, cheer-led by hacks in the pay of multinational corporations who reassure their bored readers that there is no need to rouse themselves from their stupor. Zer0 Books knows that another kind of discourse – intellectual without being academic, popular without being populist – is not only possible: it is already flourishing. Zer0 is convinced that in the unthinking, blandly consensual culture in which we live, critical and engaged theoretical reflection is more important than ever before.

If you have enjoyed this book, why not tell other readers by posting a review on your preferred book site.

Recent bestsellers from Zero Books are:

In the Dust of This Planet
Horror of Philosophy vol. 1
Eugene Thacker
In the first of a series of three books on the Horror of Philosophy, *In the Dust of This Planet* offers the genre of horror as a way of thinking about the unthinkable.
Paperback: 978-1-84694-676-9 ebook: 978-1-78099-010-1

Capitalist Realism
Is there no alternative?
Mark Fisher
An analysis of the ways in which capitalism has presented itself as the only realistic political-economic system.
Paperback: 978-1-84694-317-1 ebook: 978-1-78099-734-6

Rebel Rebel
Chris O'Leary
David Bowie: every single song. Everything you want to know, everything you didn't know.
Paperback: 978-1-78099-244-0 ebook: 978-1-78099-713-1

Cartographies of the Absolute
Alberto Toscano, Jeff Kinkle
An aesthetics of the economy for the twenty-first century.
Paperback: 978-1-78099-275-4 ebook: 978-1-78279-973-3

Malign Velocities

Accelerationism and Capitalism

Benjamin Noys

Long listed for the Bread and Roses Prize 2015, *Malign Velocities* argues against the need for speed, tracking acceleration as the symptom of the ongoing crises of capitalism.

Paperback: 978-1-78279-300-7 ebook: 978-1-78279-299-4

Meat Market

Female flesh under Capitalism

Laurie Penny

A feminist dissection of women's bodies as the fleshy fulcrum of capitalist cannibalism, whereby women are both consumers and consumed.

Paperback: 978-1-84694-521-2 ebook: 978-1-84694-782-7

Poor but Sexy

Culture Clashes in Europe East and West

Agata Pyzik

How the East stayed East and the West stayed West.

Paperback: 978-1-78099-394-2 ebook: 978-1-78099-395-9

Romeo and Juliet in Palestine

Teaching Under Occupation

Tom Sperlinger

Life in the West Bank, the nature of pedagogy and the role of a university under occupation.

Paperback: 978-1-78279-637-4 ebook: 978-1-78279-636-7

Sweetening the Pill

or How we Got Hooked on Hormonal Birth Control

Holly Grigg-Spall

Has contraception liberated or oppressed women? *Sweetening the Pill* breaks the silence on the dark side of hormonal contraception.

Paperback: 978-1-78099-607-3 ebook: 978-1-78099-608-0

Why Are We The Good Guys?

Reclaiming your Mind from the Delusions of Propaganda

David Cromwell

A provocative challenge to the standard ideology that Western power is a benevolent force in the world.

Paperback: 978-1-78099-365-2 ebook: 978-1-78099-366-9

Readers of ebooks can buy or view any of these bestsellers by clicking on the live link in the title. Most titles are published in paperback and as an ebook. Paperbacks are available in traditional bookshops. Both print and ebook formats are available online.

Find more titles and sign up to our readers' newsletter at http://www.johnhuntpublishing.com/culture-and-politics

Follow us on Facebook at https://www.facebook.com/ZeroBooks

and Twitter at https://twitter.com/Zer0Books